I0426722

500 SUDOKU PUZZLES

Medium to Hard

250 Medium Puzzles

250 Hard Puzzles

Elmsleigh Designs

TABLE OF CONTENTS

INTRODUCTION

Introducing "500 Sudoku Puzzles for Adults" – an engaging collection specially crafted for puzzle aficionados seeking a cerebral challenge. This comprehensive puzzle book is your gateway to the intricate world of sudoku, designed to cater to a wide range of players. It features 250 medium-level puzzles to build your skills, and a whopping 250 hard puzzles to truly put your logical thinking to the test.

The subsequent 250 medium-level puzzles offer an excellent bridge to enhance your abilities and build your confidence. Yet, the book's main attraction lies in the 250 challenging puzzles that will truly challenge your mental faculties. Each puzzle is meticulously designed to provide a rigorous examination of your logical skills, guaranteeing hours of engaging entertainment and cognitive development.

To support your journey, this puzzle book includes comprehensive solutions for each puzzle, serving as a valuable companion for individuals who are determined to excel at sudoku. "500 Sudoku Puzzles for Adults" offers an enthralling and fulfilling experience, allowing you to refine your mental sharpness and experience the immense satisfaction of conquering some of the most intricate sudoku puzzles available.

Sudoku # 1 - Medium

5	9		4			2		7
	2	4			5			6
								8
	5		9	7	1			
7		2					8	
9	4							
		9	2			4	6	
	6			3				
	8	3						

Sudoku # 2 - Medium

		7						1
5			7	4			2	
2	3				1			5
					2			
1	2		3					6
		4		5		1		
			1		7		5	
3							8	4
9	5			8	4		1	

Sudoku # 3 - Medium

	7		3				4	8
		8	6				2	
	9	4	2					3
2	4		7					6
		6		1	9			
			4				7	
	2	9						7
	3		5	8		2		

Sudoku # 4 - Medium

	7				1	6		
9			6			5		3
	3		2		7			
4			8					5
		3					8	
6			1			2		
			5			9		
		5		9		8	4	
1			3	8			5	

Sudoku # 5 - Medium

3				2				
		5	6		1		2	
		1	4		5			
	3				8			7
1		8						
7				5	6			9
4	5		3			8		
		9	5				4	
							5	2

Sudoku # 6 - Medium

		4		7				8
		9		3			4	5
	6				8			
				2				
5		2	9				3	
	3		1	5	4	6		
	8			6	3			
		7	5					
		3		1		2		

Sudoku # 7 - Medium

7	4		3					6
8		5						9
	9		5	4			8	
		7						
4			9		7			2
	5					6		7
		8						
5			7	3			1	
				1	4	9		

Sudoku # 8 - Medium

				6			9	2
	2		4	7	5			1
8								5
9				1	2	7		
	6						1	
	7	2		8				3
7		1				3		
			5					7
		4	1			9		

Sudoku # 9 - Medium

			9		2			
5							3	
2	1	4						
			4	6			5	9
	4			5	7		2	
		7	2	9			6	
		5			9			
9					3			8
8	6							5

Sudoku # 10 - Medium

		8	3				5	4
9		3						1
		9	8	3			4	6
	3				6			
	6			5		1	3	
		2				4		
5			4		9		2	
	7			1			8	9

Sudoku # 11 - Medium

	3							
		9						4
4				8				1
5						1		
	7		2		1		4	
		3	4				2	7
	5		9	1				
8		4			3	6		
3					2	8		5

Sudoku # 12 - Medium

							5	
8								2
	7		1	5		3	8	6
5	6		4			2		
	8	2			6			5
3	1		5		7			
6		9	2		4	8		
4							3	

Sudoku # 13 - Medium

			9		3		8	
5	9				6	1	7	
								6
						3	6	
2		7	6					
					8		5	
6			4				1	
				6	1	7		2
4		2	7					5

Sudoku # 14 - Medium

	9			4		7	3	2
7								
		2	8	7		4		5
		4		3				1
	1	8	6		4			9
		3		9		8		
								3
1								4
						5	1	

Sudoku # 15 - Medium

		2					5	
8				5	6	9		
	5		7					6
2			8		3	4		
	9				4	1		7
			1					
		3				5		
	8	9				3		
	6				8		7	4

Sudoku # 16 - Medium

							6	8
					4			2
	7			2			3	
	3	7	4			1	5	9
	1							7
			1		9			
	8	3					1	
2	9		3	1			8	4
		4				7		3

Sudoku # 17 - Medium

8		9			7			
4			8			6		2
7					1	9		3
	6						5	8
			1					
						1	2	
		4		3		2		1
				1	5			4
2	9		4				3	

Sudoku # 18 - Medium

9				8	1	6		
5							9	
8	3	6		9	4			
	6					3		
			9					7
				6			2	
		7	1		8	9		
		8	4	7				5
3	2							4

Sudoku # 19 - Medium

	5	1			4		9	
				7	9	8		
						5	3	
	3	5		9				1
	6		8	3				9
		9		2	7			
	1	6				3		2
7								8
	2		7				6	

Sudoku # 20 - Medium

	2	5	3	8				1
	1		2					7
	4	9		7	6			
			4			1		
		2					6	
			5	6	3			
9		8	6					
			7		4		5	
		4		5			2	

Sudoku # 21 - Medium

1				5	9			7
	8					6		
	7	3					2	9
				4		8	9	2
	9					1		
	2	1	3		6	7		
		9	6	2		4		
					3			
3				8			7	

Sudoku # 22 - Medium

		8		4		5		
6	4		1	3			9	
				7				
		7	4			3	5	
				5	9			
	6				1		4	9
		4		1				
	5						8	2
	2					9	3	

Sudoku # 23 - Medium

8		7	5				9	
	9			8		6		5
7	3	9		6				
				5		4		
1	4	5						
2			6	9	3		1	
	7			1				9
	5			2	8		4	

Sudoku # 24 - Medium

			6	1	9		8	
		3			4	9		2
4	7					5		6
					7	4	9	
1								
7					5	8	6	9
	9			6		1		
	8	1	2				4	

Sudoku # 25 - Medium

	2		9	8			1	
	6				3		2	
				2	5	8		4
		4		9			5	1
8								
	7	9					8	
						7		
								5
1	9	7	2		4	3		

Sudoku # 26 - Medium

4		9				2	7	
5	6			9	7			
				3			5	6
	8	4						
	2			5	3	7		
			1	2			6	
2			7			6		5
3	1			8				7

Sudoku # 27 - Medium

1	5		3					
		8					3	
		2	7		1	4		
8	3					9	2	
		1			5			
5					8			
		3				5	7	9
				1		3		6
	7					1	8	

Sudoku # 28 - Medium

	2			5	8		7	
							6	
			7	6				2
		5	2					
2			5		6	9	4	
8		7	1	4			5	3
5								1
9		6		7				
7	4							8

Sudoku # 29 - Medium

	2		4					
	9							
			3	1	2			6
			9	4		3		
7			1	6		9		5
		4	2	7	5			
	7			5		2	3	
	5					8		
8						6		9

Sudoku # 30 - Medium

		5	9	8			7	2
								6
	2	8			1			
	3			9		4		
9			3	1	5			
1		7		2				
				5	8			4
	7	9		6			8	
			7					

Sudoku # 31 - Medium

	3			7				
6	7	9		2	1	4		
	4					1		
	1		2			8	6	4
		6	1		8	2	7	
		7			9			
			3					
		4				9	3	5
				1			4	

Sudoku # 32 - Medium

			9		7	4	8	
	3			1	5	9		
		9			2	6		
			2			1		7
	6	3	7	8		5		
1	4	2	3					
9				2	8			
				5				6

Sudoku # 33 - Medium

			2	6	7			5
		3				7		2
		7		5	3			1
						2		4
7	4			9				
	2		8	4				3
9	1					3		
3					9			
8			4			6	2	

Sudoku # 34 - Medium

			9		5	6		
6				3	2		7	
3						4		1
9	7		4		6			
8	1		7					
								6
4					8	3		
	5	8		2	9			
	6				4			

Sudoku # 35 - Medium

9		2	4		3			
7		6	1	9	2			
4			3					
				8		1		
6	5	8			4			
8	4						3	
			2		5	6	7	
					7	4		1

Sudoku # 36 - Medium

								5
					5	2		
3			9	1	4		7	
9		5	2		8			
8					1			4
						1		7
	8		7			5	9	
	7	3	1	6		4	2	

Sudoku # 37 - Medium

				1			7	4
7					2			8
5					3			1
9					7		4	
	7	4	8				5	
	2			4			8	6
8				3	5			9
				9				
1	3			8		6		

Sudoku # 38 - Medium

	3				4			7
		4	9			3	5	
		7		6	5			9
	7		5		6	8	2	4
						5		
6	5			8				
	4				7			
					3			
	2				1		9	

Sudoku # 39 - Medium

4			2	7		6	8	
			1	4		3	5	7
		8	5					
	4	6		2				
	1					2	7	
			3		5			
		1				5		
	9			1	7		6	
			6					

Sudoku # 40 - Medium

							9	
7		8	4	3	1			2
4		2		7				
					4		2	
	2	1		9	5	8		
	8					9	3	
			5		6		7	9
							5	
			8	4				

Sudoku # 41 - Medium

					7	1		
			5	3				
2	3	6				5	4	
4	1	3					8	9
	2		9	8			3	
	7							
3					6		7	
		1					6	2
			2	9	3		5	

Sudoku # 42 - Medium

				1	7		4	3
			5	2		1		
	5	1					8	
	6		8			2	5	4
			4	7	2		6	
		8		6				
7		6					2	
	8			9		3		

Sudoku # 43 - Medium

	3	2		7		9	4	
6					8			
				4	9			
5			4				9	
				1	6			
		1			7		6	2
8		6					5	
3	7				4	1		6
9						8		

Sudoku # 44 - Medium

	3		2		5			
				6				2
7						8		9
	7	5					3	1
		9		1		5		
	6	2				9		
		7			9	2	1	
6			7		2			8
			1		4			

Sudoku # 45 - Medium

				3				
5	9			1				
	1		6			5	4	7
7					3			1
						2	5	
							3	9
8	7		3	6				5
				5				8
1	3			2			7	4

Sudoku # 46 - Medium

4				5	6			
						3	4	6
	3		7		8			9
5	1				2			4
	2							
			9	6		7		2
3								
			6		1	4		8
9			4			6		

Sudoku # 47 - Medium

9		7		6				
	8	1				7		9
6							3	8
	4		5	1	2		7	
					3	8	4	
		4				6	8	
		2	4	7				3
				3				2

Sudoku # 48 - Medium

5			3		6			
1		7		2			6	
			1				8	
9			4			1	7	
	5	3		1		4		
		1				6	5	
8						2		
6			8		2			1
3	1							

Sudoku # 49 - Medium

	4							2
	1	3	7	9				
	7	2		8				
						1		9
	3		6				5	
		4			5	3	8	
	6		9	5	1			3
		1				7		
			3			6		5

Sudoku # 50 - Medium

7	1		8		9	2		
					1		4	7
6					5			1
	8					6	1	
	3		9	7			8	
	7			1	8			2
3		5			4			
		8	2				5	

Sudoku # 51 - Medium

						6	7	
		6		9	8			1
		5			6	3		
	6	2	9	1				
		4					9	3
					5			6
	3							
2		8	1	6				4
4				8	9			

Sudoku # 52 - Medium

2								5
		9		3		7	1	
	7		6		9	3	2	
				2	3			
	2		5				4	
		3		4		1		
							8	1
			2			5		
8		4		9			7	

Sudoku # 53 - Medium

		2				8		1
3								
		7	3	8	9			
						9		2
	4	1		2				7
	5		9	4				
			4		2	6	9	
7					6			
	6			9	3		2	

Sudoku # 54 - Medium

8				9			7	
4			1	8				6
			6			2		
				4			2	
	7		3					8
	4				8			
	2	6	8				1	
			2	6		5	3	
1	9		7		4			

Sudoku # 55 - Medium

		2	3			6		8
	6	1			8		2	
3	8		2					7
		6					1	
					6	2	7	
			1	4				
8								
		9		8			6	4
	1		7	9	4			

Sudoku # 56 - Medium

	8						1	
9				3			7	8
								6
6	1			4	3	5		9
	4		1	9				7
5		9						
		3	9				8	1
	5	6						4
				8	7			

Sudoku # 57 - Medium

		3		9		4		2
	4							5
5				7		3	8	
	9	8	1				3	
	5				4	2		
					3	8		
	2	5		4		1	7	3
		4						6
		7	5					

Sudoku # 58 - Medium

6			2		5	8	4	7
	3	5	8	7				
			1		9			
	2	3						
			3		8		6	
7		8					9	
1							3	
	9		6	4				
		7			1	6		5

Sudoku # 59 - Medium

7					4	5		3
					3		7	
		1		8		2		4
		2	3				5	
			2			4		1
		5		1				2
	9							7
3			6	9				
6			8		1		3	

Sudoku # 60 - Medium

7	9	4						
		3	7	6				
5								
		7	8		6	1	2	3
			1					4
				3	4		6	
	5					7		1
6				1	8			
					7	3		6

Sudoku # 61 - Medium

4			7	8			3	
		8		6			1	
		3			5			
9	8			3	2		5	
2		5			8			9
	4		9				8	6
								3
						1		5
	3	2			7		6	

Sudoku # 62 - Medium

	3			9		2		6
		9						5
6		1				7		
7			1					
9				2	4	3		7
	5	4						
	4		5		7			2
8								4
5	6		8		9	1		

Sudoku # 63 - Medium

		3				7		
2	5		7		3	6		
					4		9	
9		4		8			2	
				9				
7		8	3	2			6	
			2	7		8		6
4		2	5					
								2

Sudoku # 64 - Medium

				6	7	9		
5	8	6	1	9	2			
						2		
3					6	4		
			2	7		1		
			3				2	
	6	2			5		4	7
	5							
	4				9	6		

Sudoku # 65 - Medium

			9		8			7
	8	9			3			4
	2		1					9
			3			4		
6				1			3	
9	7						5	
				4			6	
1		5	2	3		7	9	8
3								

Sudoku # 66 - Medium

	6	3			2	8	5	
		7						
			6	4			3	
	7	8		9			6	3
			1			5	7	
		2				9		4
6								5
	3				4	6		
		1	2			3		

Sudoku # 67 - Medium

1	5							
8	3			9		1	4	
				5	1	6		
			2				9	
	4		6			5		
	6		3		5		2	1
						8		
		1			3	2		4
4				7				

Sudoku # 68 - Medium

	6			5		9		
7	4	2		8		3		1
	9		4					6
		4		3		8		
		9				1		
	3		9		8	7		
					4		9	
					2	6	8	3
	8							

Sudoku # 69 - Medium

		7			9			8
4		3			2	6	7	
		2	8	7				
					3	5		
		9				3	8	
					5		9	7
	3	5					6	
7		6	4					
1						8		5

Sudoku # 70 - Medium

	7				8	2	4	
1					6			9
		9	3		2	1	5	
			1				6	
9						7		1
8			2					4
						9	3	
				2				
		1	5		9	6		8

Sudoku # 71 - Medium

5	9		1			6		3
	2						5	9
	3		2		9	8		
		8						
						1		
	1	5	9	3	6			
6				4			8	
		3		8	7			6
						5		

Sudoku # 72 - Medium

1							3	
	3	9					6	
7				6				2
			5				4	
5	2		1		8			
3	6		4					9
		5	7	3		8		4
			2		1		5	3

Sudoku # 73 - Medium

		8				2		9
					9		4	
4			1			8	5	3
		1						
2		7	5	4			9	
9			8				1	
6	3	4						
8	5		9	6				4
								6

Sudoku # 74 - Medium

	2	9		5		1	7	
7		6		3				
				2	6			
5		4		7		6		
9			5				1	
		3	1	9		4		
6				1				9
						7		1
	9					3	2	

Sudoku # 75 - Medium

	1				3			
8		5			6			
4	3	9	2				5	
	4		3		9			
3		8			1	4		7
				2			9	3
1	5	4	9					
9			6	8		5		

Sudoku # 76 - Medium

5					1			2
			4					1
				3	2	4	8	
2	1	4				5		
		5					6	
6						1		
4	2				8		9	
	7			2			1	4
		8	6	1				

Sudoku # 77 - Medium

3		4				1	7	
1		8			2			3
	9							4
						2	1	
				8	5	7		
		9						
	3		4					
9	4		2	7			5	
6	2				8		9	1

Sudoku # 78 - Medium

3			6		8	7		
	4							6
		2		3				4
1								7
	3	6		8				
	9		5	2			3	8
						5		9
8	2				5		7	
	6		3					

Sudoku # 79 - Medium

			3	1	5			
9	4				6			5
				7			2	6
						7		2
2		8				5		
				4			8	1
	3		2	6				
			9		1	2		
8		1	7			3		

Sudoku # 80 - Medium

		4	5		2	1		8
	7							
8						2	3	
		9		6				
			2	4				3
3	6			1				
			4		1			6
	2				7	3		
		5		3	9			1

Sudoku # 81 - Medium

			1			7		6
8			9					
				7	3			1
								9
	5	1			9	4		
	9	6	5	8				
		9		3		6		2
	4			5		3		
			2		4	8		

Sudoku # 82 - Medium

					2			8
	2		5					3
				4				
7		5	9		4	8		
				7		6	5	
	4					7		
5	9				1			
4		1	3	9				
	8		6	2			1	9

Sudoku # 83 - Medium

		3		7	2	6		1
				8		5		
2	9						8	
			9				1	
		6	1					
1				2	8			
					4	7	6	
4		7		1				9
6	5	8	2					

Sudoku # 84 - Medium

9	6							7
5	4	3	1	6		9		
		8						
		9			6		8	
	5	4	7	8				
	1			3	9		6	
			8	7	2			4
			9					6
			6	5				

Sudoku # 85 - Medium

		6		9		3	7	
	4	5	8	6			9	1
						5		
4			1	2			8	
		3	9					
5	1							
		9		1				2
	5		3					
	6		5	8	9			

Sudoku # 86 - Medium

						4	9	
	7		5					2
	8		4					
	9		8					7
		2		7		5	8	
					5	2	6	
								6
8		4	7	1				3
6				8	3		1	

Sudoku # 87 - Medium

		7	8	9	5			4
	5						6	
		2						
2				4				7
	6	4		1	3			2
					2		3	
9				3				
	4		1	7		6		
					6	9	4	3

Sudoku # 88 - Medium

	4		8					
	7			3	9			
9						8		
		8			6			2
5	3				2			
7		2			1			6
		7	9			2	5	
			1	2			7	
	8			7				1

Sudoku # 89 - Medium

		5	3			9		
4					9	8		5
	9	8	1					
6	1						8	
9				7		6		
	8				6			
	4						3	
				1				6
7		6	2		4		9	

Sudoku # 90 - Medium

2				1		9		
			5	9	7		6	
			2	6				4
	8		6	7	2		3	
	7		4	3			8	5
	1	3			5	7		
	2				6			1
							7	
			8					

Sudoku # 91 - Medium

2		8		3		5		
		7			4			
3			1			7		
				5	2			8
	2					4		
8	7					1		
6		2	9					
	9	4	8	6	1		5	3
			2					4

Sudoku # 92 - Medium

5			3					8
			5	9		4	6	
1		2	8		6			
8	3							
7	1		6				4	
6			9		7	1		5
		3				7		
					5	8	2	
		7					5	

Sudoku # 93 - Medium

3	2			6				
7			9		2			
			7			8		
5								1
6			2	5	8		4	
			6	7	1			
		1	8	2				6
9		7				4		
				1		9	3	

Sudoku # 94 - Medium

				7				
	3			9		1		2
6		1						8
			7	8	9	4		
3			4			2		
5				2		7		
4						6		
2	6	3	1			9	4	
		5						

Sudoku # 95 - Medium

	2		1					
7							6	
		5		4				
2		8						
	6		3				4	5
3		7			6	8	9	2
		2			1	9	3	
	9			6				
4				9	2		8	

Sudoku # 96 - Medium

7	6				2			5
			7					
5								
		6					8	
9	5	8	6	1		3		
				4	3		6	9
4				6		9	1	8
8			2		4			
	9	5						4

Sudoku # 97 - Medium

4	5	9						2
		7		2				
					6			1
		4	6					
5				3				6
		2	1		4			3
1					2	5		
					3		7	8
7		8	5				2	

Sudoku # 98 - Medium

			7		6			
2	3				8			4
	7	9				1		2
3	9				4	2		7
							3	
4	1						9	
					2	9		3
	4	1						8
9			5			7		

Sudoku # 99 - Medium

2	7				5			
			2	4		3		
9	3				1			
6			5			9		
						1		4
	9				6	8		
8		6	1	3				
7	2	3			4			
4			6			5	2	

Sudoku # 100 - Medium

2						1		
8		4	3				6	5
	3	1		9				4
5								7
7								
		9		7	5	2		
4				1	6			
3				4	8			2
						7		6

Sudoku # 101 - Medium

	1		8					
6	9			1			5	
					2		6	
		1	6				3	4
5		6				2		8
		3					9	
			7	9		4	1	
	6			2				3
					8	9		

Sudoku # 102 - Medium

		5			1			
		3					4	
8			2		9	7	1	
1			6					
	4				3			
3	9					6		8
						3		9
5	3				7			
		9	4	3	2		5	

Sudoku # 103 - Medium

	1		5		7		4	
5	4		9	1		6	7	
						5		
	8	6				1		
	3			8			2	
		9	7			4		
6			2					
8					5		6	
		4	6		3			

Sudoku # 104 - Medium

1		9				2		
	2						7	
5		4		3	2		1	
				5	1			3
		8	2		3	1		6
	1			9				
9	5							
8		2			5	4	9	7
								1

Sudoku # 105 - Medium

		3	9	2			1	6
	5	2	8	1			9	
		1						4
		4			1			
				6				8
		7		9				
4		9	5	7			8	
		8			9		6	
3					8	9		

Sudoku # 106 - Medium

5			4			2		
7			5	3			1	
	9							
1	8	7		5		4		
		9				5		1
			6					7
			3	7				8
					9			4
8		6		1		9	5	

Sudoku # 107 - Medium

							7	
	8				1	4		
4		9	3			5		
				3	9			7
	7			2		1	5	3
		2		1		8		9
	4			7			9	
	2				5	7		
5					2			

Sudoku # 108 - Medium

	1			8		6		
		5	1			3	7	
4							2	1
	2		3	4			5	
		8				7		
		1						
	5	7		1				6
	9		5			4	3	
	8		2					

Sudoku # 109 - Medium

6			2					3
	2		6		1		9	
3					9		8	6
					6			7
7	1			4				
			1	2		8		
2	7		4	1			3	
1	3							8
		6			8			

Sudoku # 110 - Medium

		7	5	2				
	4	2		1		6		
1							2	
9	8		3				1	4
	7	1						5
			4		1			
6						8		9
		9	2			1		
				4	5			

Sudoku # 111 - Medium

2	6	9			5		7	
			8		7			
8	3					4		
4			5	7	2		9	
					8			4
7		2			1			
	5							7
			1					
		1			4	8	5	

Sudoku # 112 - Medium

	6			2		8		
2		8		6				
							6	
	1				3			5
6	8						3	
		3		4	9			6
8		9	1				4	
			4			9		1
	5			9	7			

Sudoku # 113 - Medium

	6					9	7	
								8
	4	7			9	3		
			6	5		1		
					4		2	
	9	6						
		4		7				3
3		8			2	7		6
		2		9	3	8	1	

Sudoku # 114 - Medium

	2		6			3		5
7			3	5			4	
		5		2				1
6		8	9				5	
		3		6			7	
							2	
		1		3				9
			2	1				4
5					8			

Sudoku # 115 - Medium

	8			5				9
			2	4			1	
	1	9	6		3			
3			9		6			
				7				5
	2						6	
		1	5	9			7	
		4	1			9		8
	9	7						4

Sudoku # 116 - Medium

3			9	7				5
					6			
	7		5	1				
7		6			9		8	2
5			7					6
		9				7		
	6							4
				6		3	1	7
	4	3	1	2	7			

Sudoku # 117 - Medium

9				4			6	3
4						9		
1		3	6			7		
		4	2		8			
					6			9
			3			2		4
5		6		1	7			2
2	3					1		
				3		5		

Sudoku # 118 - Medium

		2				5		
					4			
		5				4	6	
	2	3			7	8	5	6
	1	6	3					
5		8	6		9		3	
3				7		9		
						1		3
					1	6	4	7

Sudoku # 119 - Medium

					2			4
		9			1			
	5		6	7			2	
1		2					4	
		5						8
9				2	5		7	
6	2		7		4			1
			2		9			7
7		8		3	6			

Sudoku # 120 - Medium

6	1		8		7	5		3
	3			5			8	
					6	9		4
		7		3				
				7				
		3	4		5	2	6	
3	2			4				
	5	9		8				
					2	3	5	

Sudoku # 121 - Medium

	6						2	5
						7		
1			3	8	7	9		
						1	6	
4		2		6				
8								4
		3	2	5			7	1
	5	8			6			
	4			7	3	6		

Sudoku # 122 - Medium

		2					1	7
		1	3			9		8
					8	6	3	
9							8	
	5	8	9	3		1		6
	6		1					5
								9
1				5	3		2	
			8		4		6	

Sudoku # 123 - Medium

	4		5				6	8
7						2	3	
5			6	2		9		
				6				
			1		7	6		
	1	6		8			2	7
		9					1	
2				1	8			
			2				8	

Sudoku # 124 - Medium

			9	6		5		
						6		4
	3	5			1			8
1						3	8	
7						2		
		9	1		3		6	
		6	3	9				2
2	7	8	4					
		4		8				

Sudoku # 125 - Medium

1		6			8	9		
			2		7			
7	8		6		9			
		1						
			1			2	3	
6		5	7	9	3		1	
		9			5	8		
2	4						9	
					2	7		

Sudoku # 126 - Medium

9	6			5				
		2						
				9			4	6
8	9		2			4	6	7
						8		
7	3		8					
3						5		1
6		9		1	5		3	
					3		2	

Sudoku # 127 - Medium

	5	6			4			1
	9				6			
						7		
5		2		6				8
9		8	3			1		
4	7						3	
8					3			5
		5		2			8	
			7		8	3		

Sudoku # 128 - Medium

				8	6	3	2	
2	6					5	8	7
	8				1			
		5				4		
				1		7	3	
				4	3	2	9	5
6		9						
8					2			4
				5		1		

Sudoku # 129 - Medium

3			5	4		9	7	
1		4		8	2			6
2			6					4
6						1	2	
		9				6		
		7			4		3	
9		6	2			7		
		1	8					

Sudoku # 130 - Medium

						4		8
		6		8				2
4					6		7	5
3	4		2					7
5					1			
			4	3				
		8		4	9			3
9			3			8	5	6
			8			2		

Sudoku # 131 - Medium

						9		3
			5		7			
				4		8		
	7				6			
8	9	2				7		6
6			7			2		
		7		3				4
	8	4			2		5	
2	6		1				9	

Sudoku # 132 - Medium

5					8			
6		4						
8		7		2		5		
3			5				1	2
		2	3			8	5	7
				4			6	
2		1						
	4		1		9		2	
			8	7		6		

Sudoku # 133 - Medium

	7			8	9			
2					1		6	
	8			6		4		
	2						3	
		4						9
5				3				6
8			3	4		6	1	
			6		8			
6		2	1			7		3

Sudoku # 134 - Medium

		1			5			
	3		7	8				2
8								7
	6	2		4		5		
4		3		5	8			
5						6		
							2	4
	4			3		8	6	
1		8	2					

Sudoku # 135 - Medium

		3		4	5	9	7	
					1		2	
1			9	2		8		4
3	1					6		7
							1	
9	5							
			7		9	2		
	3		8				5	
8		7						

Sudoku # 136 - Medium

	7	4				8		2
					4		6	9
	8		9	1	2			
	6		3		9		5	
		1						4
2						3		
	3		7		6		4	
			2				9	
	1					7		3

Sudoku # 137 - Medium

	3		4	9	5	7		
7								3
	5					2	8	4
		1		3				
2			5				7	
3	9				2	1		
	4		7				3	
								2
	2			4			9	

Sudoku # 138 - Medium

5	8	4						
			8	4		3		
			6					
			4		1	6	8	
8	1		2	5			9	3
								7
	9			8	4			
	3	8	7	1			5	
					3		1	4

Sudoku # 139 - Medium

	2						1	5
7		3		1			6	
			6					
	5	2	9	4			8	
	4				8			6
					7		2	
					9		4	
			4	8	6		7	9
4					3	8		

Sudoku # 140 - Medium

					7			2
	1	9	6				8	5
				5			6	
	7	8	5					9
						8	7	
		3	7	9				
	6	7				2	4	
2				6	1			
	8	5				3		6

Sudoku # 141 - Medium

				6	8	1	3	
3							9	6
9		6				5		
	5		3				7	
		7		8				
				5	1	4	6	
	6	1	5				8	
		9	6					
		4		2				9

Sudoku # 142 - Medium

				7			6	
		2				8		5
5			3	8				
					2	9		
			6				3	
3		5			1			2
	5	7		6			2	3
6				2				
	9	4				5	7	

Sudoku # 143 - Medium

	8			3	2		1	
3			8					
				1		5		
		8	3		9	1		
5		4						
9	1			4		6		
			9	7	3	2		
	2	7						9
1	3			6				

Sudoku # 144 - Medium

8		6		3		9		7
1				2				
	4		1	6	7		2	
3			2	5			7	4
						1	5	
			9					
9					8	3		
	3		7					
		1	3					

Sudoku # 145 - Medium

5	6				2			
			5	4				8
						2		4
3	4		2	6		8		5
					8			
	1	8				4		
			8	1				9
			9	2		6		1
1		7			4			

Sudoku # 146 - Medium

				4			6	
	8			3			7	
	2	5				9		3
			2				5	
7		2		5	8			
		9	7		3	4		
	1				9	7		
3		8	1			2		
2		4						

Sudoku # 147 - Medium

9	2			8			4	1
1			5			6		
						5		9
	1		8				6	
								3
		2		1		7		8
	7	9			6		3	
	6							4
4	5	8		7		9		

Sudoku # 148 - Medium

							1	
7	5			3				
8	6				4			5
	7		8		3		5	
9	2					8		
	1		4			3		9
	3				2			7
		9			8			1
	8			5		4		

Sudoku # 149 - Medium

	2		8	1		7		
		6					5	4
7								
1				4	3			
	6	4	1					5
			9		6	4		
	7	1	5		9		4	
	3		4				8	
	5						3	7

Sudoku # 150 - Medium

	3			5				8
9		2					7	
	7		3	2	6			
8						9	2	
			9	8	2		4	
6			1				5	7
4					3			
	9	7						
					1			9

Sudoku # 151 - Medium

2		4			6			
		5		1	7			6
		3		5	8			1
		7		8	2			4
		1		4	3	5	7	
					1			
					5		3	
		8	1					
				7		1	2	5

Sudoku # 152 - Medium

1	2							
		8			9		6	
				2				
6	3			5				
			7			3		
		2			1		7	6
5		6		4	2			7
			3	9	7		1	5
7	1		8					9

Sudoku # 153 - Medium

	4		5				8	3
			8		9	6	7	
3			7					
4	2			8		5		
5		7			1	2		6
		1						
	1		2		8			
	7							5
			4				2	

Sudoku # 154 - Medium

4	2		8				9	
					7	8	4	
	6		5	4			1	2
		1						
				3			7	
			4	5		6		
	3			2				5
			3			1	8	
8		5	9					

Sudoku # 155 - Medium

1			2	7	4			9
			5			3	1	
2		8						
8	2	7			3			4
	6			2		7		
		4		5	1		3	
				9			5	7
					7			6

Sudoku # 156 - Medium

	2		3	4			7	
	1	9	5				8	3
		2			4	6	5	
8	6			7				
	3	5		2				7
			4			9		1
		7						
3	5				6			

Sudoku # 157 - Medium

4	2			9		5	1	3
					8			7
		9		1	3	4	8	
			9			7		
8		1	2		4		5	
5	9	4						
				3				
1				7	2		3	
								6

Sudoku # 158 - Medium

9						6		
		3		4			2	
		4	1	5	8			
6	5		3	8				
		9			1		3	2
3					9			8
			8	6	2			1
		8					9	6
		6						

Sudoku # 159 - Medium

	1			2	7	8		
9					3			
	3		5		9			4
8					2	9	5	
		9	3	6		4		
7							6	
			8	7		2	4	
	8		9				1	
	7				1			

Sudoku # 160 - Medium

		1			9			4
		5		1		3		
		3		6		2		
	5				8			7
1		8		9	7			
4	7			5				2
2	1			8				
5								3
8			5	2			9	

Sudoku # 161 - Medium

						4		
			2	5				9
	9	8	6	7	1			2
	5							
		7					4	3
		1		2	7			
					5	6	9	
1					6	2		8
	4		3			7	1	

Sudoku # 162 - Medium

			4		2	6		7
				9	8	3	4	
		9		1			2	8
6		2			4			
1							9	
8		1	2		6			5
			7					
		6		3	9		8	

Sudoku # 163 - Medium

					2			9
	3	1				2		
9		7		5			6	3
			1		7			
		9	6	3				7
	7				5		2	6
7	8					3	9	1
					8	6		
	9						8	

Sudoku # 164 - Medium

7							5	
		5			9		1	7
6		1		4				8
			8		1	5		
8		3			6	9		2
	7						6	
	2							
	4		9	3		7		
9			7		2	3		

Sudoku # 165 - Medium

		8			5			
		7			3	6		
		1				7	2	5
			6		2			3
			1		8		9	
		3				4		
	8	6	3			2		
9		5		7				8
		2	4	8				7

Sudoku # 166 - Medium

	6	8	2				5	
						9		
		3			6	4	1	
		9		8				6
				5	4			
			9			8		
			4			1		
2	7		5		8			3
8	5		6		3		7	

Sudoku # 167 - Medium

6	1					8		
			8				7	1
			6	9				
	5	6			9	4	8	
4	3		5					
				4		7		
2				7	5			
		1			2	9		6
			9		8		1	

Sudoku # 168 - Medium

	6					3		5
7			8					6
1	8		5					2
5		7				4		
		3		7	6	2		
			3	1				
3					4			8
								9
2				3		6		4

Sudoku # 169 - Medium

6	8	1			7			
							5	
	3			6	8	7		
	1					9		2
				7				5
		3	2			8	4	
	6			5			8	3
	7		4					6
		2	6			4		

Sudoku # 170 - Medium

1				4				
9	8		5			6	7	
			1				4	
		9		6				3
3		1	4	8				
	5							
		2			7		8	
8	9	3						
6	4			9			3	

Sudoku # 171 - Medium

	3			4			6	
6			2			3		
8	7				6			
		1	9					3
	9	8	1	5	3		7	
					2	5		
5		3			7	8		
		6				7		
								9

Sudoku # 172 - Medium

3							6	
		2	3		1			
			2	7				9
6	4	3				2	8	
	5		4	2		9		
		7	9	6	2	4	1	5
		6		5				
							9	3

Sudoku # 173 - Medium

5				4		1	2	
1	8		9	5				4
				6	7	8	3	
		5			1			
6		9						
8		2		7	3	5		
	7		2					
	9					2		3
					6		1	

Sudoku # 174 - Medium

			8	4				
			9		2			
		2	1			7	9	
	2	8	4	7				
				2				
7	4	1	5	9				8
2	8				9			
		9		3			8	1
				8		3	2	

Sudoku # 175 - Medium

			4					
	7						3	2
6	5		7	1		4		
		7	6		8			
9		4						5
		3		9		2		
			1	6			4	9
			9	8	5		2	
	8			7	4			

Sudoku # 176 - Medium

5				2	7	1	3	
2				8				5
			4	5				9
			8	4				
	5	8			6	9		
	1	7				5		
								3
	6		7			2		
7		3			4			

Sudoku # 177 - Medium

	7	2	8			4		
1			5				9	7
9		6					8	
	2		9	3				
	9		2	5		7		3
6		7			1			
4						6		
	1				7			8

Sudoku # 178 - Medium

			5					1
2		3		1	7			6
	7		4	2			8	
5			8					7
					5		9	
9						3	2	
8	9							
7		2				1		
	4		7		9			

Sudoku # 179 - Medium

4					3			
	5		7					3
1			2				8	
			9	7				
					8	6		9
6	9	4	1			5		
9	1	2		8	7			
		6					3	
	8					9		

Sudoku # 180 - Medium

			8		1	2		
	4				9	1		
5							6	
	1	5			3			4
3	2			9				
4	9				2		5	8
7			9					
				5				
1			3		8		4	

Sudoku # 181 - Medium

		5			9	2		6
1			5				4	
2					4		1	9
8				1				
		4			2	9	3	5
	7							
3	8							1
		6		5		3	9	
9				6				7

Sudoku # 182 - Medium

	1	5						
		4		1			6	8
2		8				9		
5			9		7			6
6							4	
	7		8				1	
			2	9		7		
	9			7			8	2
		7						5

Sudoku # 183 - Medium

				7		1	6	2
								7
	7		1			3	4	
			4	5	2			
					9			
	9	5			3	2	1	
		8	6		4			
			2				7	
	1	6		9				8

Sudoku # 184 - Medium

			9	5				
				3	1	8		
9	4		2				5	3
3	5			2				
7		9	3	1			2	
	8						6	
		6	1				3	
	2			6				8
	1					5		6

Sudoku # 185 - Medium

4						5		
3		7	8					
	1	8			9			7
	3						7	
5				8		1		4
7		9			2	3	5	
			9			2		
			5			7		
9			6		7			

Sudoku # 186 - Medium

					4			7
6				9			2	3
				2		4	5	
		3			9	6		4
		9		4				
	5	1			7	2		
3					8			
	4	2				1		9
				1	2			6

Sudoku # 187 - Medium

5	1	4		9				
		7						
		6	3		1	5	4	9
8						1	6	
1		5	8	3	9			
						9		
6	7							
	3			5				
			4			7	8	

Sudoku # 188 - Medium

	5				9			2
						5		
7					1		4	
						6	3	8
		6		1			9	
		4		6	3	2	5	
8			5			4	7	6
5			1				2	
	4					9		

Sudoku # 189 - Medium

2		3				6		
	4	6					7	
				9	6			8
3			5			7		1
	7				4		5	3
6						4		
				2	8			6
1				3		8		
7		8	6			9		

Sudoku # 190 - Medium

	8						6	4
				6	7			
3			8	2				1
	9			4		7		
6		3	5					
		2		3	8			
8						1		
	3		4		9	5	2	
9		7					8	3

Sudoku # 191 - Medium

	2			3	1		7	4
		9						2
	1	4					5	
5	8						1	
				8	6			
		6	1		5	9		
	7		2	5				
						4	9	
9		1			3		2	

Sudoku # 192 - Medium

9			1			8		4
			7			2		6
		2						
					2			9
2				6		7		8
3			9					
7			4	3	5			
	2	9		8	1			7
6		5			7		8	

Sudoku # 193 - Medium

	3		6				5	
2								8
					7			
9				4				
	2			9		8	3	
			2					9
6	7		3	8	9		1	4
3	1			5				
8				7		3	6	

Sudoku # 194 - Medium

			1	5	9	7		6
			4			2	9	3
							5	
	4	3				8		1
		9					4	
2	8		6	1				
8			5					
	1		8				7	
9		6		2				

Sudoku # 195 - Medium

	8		3	4		7		
1				6	2	4		
5	3							
	6			9				3
7								8
	1	9			3		4	
	7	1	6	3				
		6			7			
			5			1		

Sudoku # 196 - Medium

	5	7		4				2
6		1	3				5	
			7	5				
								8
				1		5	6	
9		5						
		4	1		2	3		
	7		8					5
	8			9			4	1

Sudoku # 197 - Medium

						3		1
	3	8	7					9
		7						4
				2			9	
9				8	1			
7	2				4	8		6
6	1	5	4					7
					2			8
		9					3	

Sudoku # 198 - Medium

		1	9				2	
	4				1			5
			3		2			6
								2
	6	5		2			4	
			7		4		6	9
		4		9				
	1	6		7			3	8
2			8					

Sudoku # 199 - Medium

				6	2		5	
	8		5				2	
			8			4		1
			7	4				8
		3	6	5		9		4
			3		9		6	
	1			3				
3		2						
6	4	9		1	5			

Sudoku # 200 - Medium

8	5		9	6	3			4
4		7			1			3
	6		2			8	5	1
		4		5				
	9		4					
				7		1		
								8
		6				3		
2			6	1				5

Sudoku # 201 - Medium

	8	5		1		4		
	2	7			9	3		8
				8				
	3		6				9	
			1				3	
	1		4					7
	4	6	9		1			
						8		3
5			8	4				

Sudoku # 202 - Medium

5					2	8		9
	1			6	9	2	5	
							7	
		3	7				4	
							3	
	8	6			1	7		
		9			5	6		
6		5	9					
8				2		5	9	7

Sudoku # 203 - Medium

	8	1		5	9	3		7
	4		7					
		5			2			
				6		7		4
			2		1			3
	5	2		3				8
	9		3			8	6	
		7	8					
		8						9

Sudoku # 204 - Medium

			9		8		7	6
4				1				
	2			3	5		8	
							3	5
3		7						
9		8	3					7
			8		1			
			6			5		
8	9	4		5			1	

Sudoku # 205 - Medium

	7			8			9	1
	1			7				
9			6	3			2	4
			1		2			
7					3	2	4	
	2	3						5
	3				6			
				1	5	8		6
4			3					

Sudoku # 206 - Medium

5					6			
3	8		7				9	
				9				
				7		2		
4		2			9	8		3
9		8	4					
6		1					8	
		4		3	1	9	5	
				6	5			1

Sudoku # 207 - Medium

			6		9	3		5
	5						8	7
		6	2	5		9		
8			1		6	5		4
	6	4	5		2			8
9				7				
		7						6
					5			
		3		2				

Sudoku # 208 - Medium

6		2				1	7	
				9	7	2		
	7		5		2		9	3
	6			4				9
8		4				6	3	
9			6		1	4		
				1	4			
			7					
1	5	3						

Sudoku # 209 - Medium

	7						9	
	4	5		3		2	6	
3					8			
	6				7			
		4			5	6	3	1
						7		
				8	3		1	6
		1	2	6		9	5	
4				7				

Sudoku # 210 - Medium

	1		8					9
7					2		6	
8		2	3		9			
2		5				3		
		1						8
					3		2	
9	6			3				2
3	2			7		1	8	
	5			2		4		

Sudoku # 211 - Medium

	1			7		4		
2			9	1				
		6			2	1		
		2		6		5		
8			1			3	2	
4		3				9		
	8			2				5
				5		7	1	
					7			3

Sudoku # 212 - Medium

3					4	2		
	7	2	1		8			4
4				3	2	1		
8				6	5			
5		7						
		6	3			8		5
					1			
	4			5				9
7	1							3

Sudoku # 213 - Medium

2		9	5		8			7
5				6		2	9	
6			9	4		5		
				7				8
			4				6	3
			2	9			7	
8					9		5	2
	2			5				
		3	6					

Sudoku # 214 - Medium

6	1	4					5	2
5					2	6		
	3				5	8		
3			5		1			8
			3					5
	6			7				3
			7		8	3		
			9		4		7	6
4	9							

Sudoku # 215 - Medium

		7		2	5	3		
8		4		3	1	5		
5		1					2	8
					8			
1		8				4		
6			5			7		
					9		3	7
			3					
4				8	7	9		6

Sudoku # 216 - Medium

7	8		9			4		
			3		5	2		
9	3		4	8				
6		3	1					
						1		
4					9		6	
	6	7				9		
		8		9		7	4	
					2		5	

Sudoku # 217 - Medium

2		5		6		8		
9			2	5	3		1	
			7					
	3				8	7		
	4	8	6	3				
3				9		2		8
	9	7					5	
8			5				7	

Sudoku # 218 - Medium

	7		4			3		
	3				9			
2							7	
	2	1			8		4	7
		7			4	6	3	
4	9					5	8	
	6						1	
	5			3		8		
8		2		7				3

Sudoku # 219 - Medium

	7						2	5
5	3					7		
2					7	9		
9	8			7				
6	1	3		8	2			
	5	7			6	3		
	2			9				8
3				5				
							3	7

Sudoku # 220 - Medium

	7			5	6	4		9
							5	
	6			3			2	
				2				4
		8		9		3		
		2		8	1	5		
2	9	4	3				6	
		7		6				1
	5				7		3	

Sudoku # 221 - Medium

4					7		5	
		3	4			8		
7				8		1	4	2
						3	9	
		8	6		3	2	1	
		2	7			5		8
	3				1	7		
	4						3	1
			8					

Sudoku # 222 - Medium

					1			
			6	4			5	
		3	5		7	9		
	6			9		1		
				1	2	6	4	9
	4				6			5
7		1		5				
2	8						9	
5					3			2

Sudoku # 223 - Medium

8		3			2	6	5	9
								3
					1			
3	6		7					8
5	8			2			9	
		1	5			7		2
		6	3		9	2		
		8		1				7
	2			8				

Sudoku # 224 - Medium

				5	7			4
			3			9	1	7
	7				1	8		
7		8			4			3
6			7					
		3		9				
8		9	6	3		4		
	6							
	2			8				9

Sudoku # 225 - Medium

	8				4	6		2
	6				5	7		
			1		2			
		1		9			6	4
			2			5		1
	2		4		6	8	3	
2								8
9	5							
	7	8	3	2				

Sudoku # 226 - Medium

				1		4	2	
		1		5	2		3	
		7		3		8		
		3					9	
	9		7			3		
	1			6		7	5	8
		8		2	3	5		
	3							1
						2	8	

Sudoku # 227 - Medium

				7	3			8
	2							7
			2		5	4		
		3		5			6	
5	6					9	2	
	8				9			5
2		8		3	6	5		
1	5			2	4		7	

Sudoku # 228 - Medium

				7	9	8		
2								
		8			3		5	
	2	7						6
3		1					7	
6								
9		2	3	4	5		8	
	1		9			2		
8	3			2	1	7		9

Sudoku # 229 - Medium

	1		7			5		
	2			9		1	3	
			3		1	8		
3							8	6
2						9	1	
7	9			5		2		
					6		9	
		7	1					
				2		6		1

Sudoku # 230 - Medium

		2		9		5	1	
			5				8	
3	8				2	6		
	3	8	2					
		6						9
	4		9		6			
4	1					9		
	6			7	4		5	2
	2		6					8

Sudoku # 231 - Medium

8					2		7	3
	4		7	9			6	
			8		5			
3								2
6			3				5	
	5	9			4		1	
	9				1	4	3	
2				3	8			5
		6						

Sudoku # 232 - Medium

7		1			9		5	
2				4				
			6			8	2	7
					4		6	
			8	6		4		
	1			7				5
		7				2		
	8				7	9	1	6
	2	6			8			

Sudoku # 233 - Medium

9		1	7		2			
7						2		5
	2		6	8				9
4	9			3				2
					7	5		
3		2			1		4	
			3			7	6	
		5						1
1			4					

Sudoku # 234 - Medium

	8			6	3		5	4
4	3	5						
		6	8					
								5
2				9				7
		7			6	2	9	
		1	9		8	5	4	
		8			4		2	
				5			6	

Sudoku # 235 - Medium

		5				9	4	1
	3		4	8			5	2
			5	1				6
4	5		1				6	
9		2			4			
3			7					
7		3				8		
				4		6		
	9	1						3

Sudoku # 236 - Medium

	5			1	4			
4								2
9	3		2					
5					3		8	
		6	5					7
2	7				1			4
8	6			4		1		9
					6			
			1	2			3	5

Sudoku # 237 - Medium

1	9				6	4		
		4	1		8			6
				3		9	7	
		2	5					
	1							
			8	4	1			2
	8		9		4		5	3
7			2			8		
					5			4

Sudoku # 238 - Medium

			8			1	7	
5		9	2					
	6	1		3				9
		5			4			7
		4	5				9	
			1	9		5		
	4			5		6	1	3
6		3						
	5					9		4

Sudoku # 239 - Medium

			2					
	7	1		6		3	2	
	6		1				7	
		7	5				9	8
	2							
4	5		6		1			
1			9	2			8	3
							6	5
3			8		5			

Sudoku # 240 - Medium

3		5	6			9		8
	1						7	
		7				4		
5		8	3		7			9
	7		8	6		5		
		6						
			9	3	5	2		
							1	5
		4			2		9	

Sudoku # 241 - Medium

			2				9	
	1							
5				9	1	6		
3							8	9
		6	1	7				4
	8	7						
				1	2			7
		9	7				6	2
			9		4		1	8

Sudoku # 242 - Medium

		8			2		7	
	7	5		6				8
	1				3	2		9
		1	8				9	
		7			1	5		2
	2			7				
	9	2					1	3
4				1	9		2	
	8							

Sudoku # 243 - Medium

6	5	4			1			
9	8						6	5
				4				
	6	8			2	9	1	
	9			1				7
1		2	3					
		6		9	5	2		1
				2	3	5		
	2							

Sudoku # 244 - Medium

					4		1	
3		5					4	
	8							6
1	5		8		9	2		
7						6		
	2		5	3				1
	4		2			3		5
				9	3			
			1		5		6	

Sudoku # 245 - Medium

2				1	6			
	9	8	5	2	4	3		
			8			6		
8		6		9			4	
		2			5	7		
4		7				5		
	2	5			9			
			3				5	2
7			4					

Sudoku # 246 - Medium

2	1			4				
4		6		8		2		
	8		2	5			7	
			6	1				
		2						5
			8	2	7		4	1
						9	1	
9				7				6
	7					8	3	

Sudoku # 247 - Medium

4			7					8
	3					2	1	
	5		2	4	8	7		3
2	4	9						
		3	9		5	1		
		5		8			7	
					6			9
5	9					3		7

Sudoku # 248 - Medium

7				4	1			
		6			9		3	
		8				6	4	
	7			1				
8	2							4
	6				7		8	
				9			7	8
4				7	3	1		9
	9		1	8			5	

Sudoku # 249 - Medium

				2			9	4
4	8			3		1		6
9								8
							7	
						9		2
			6	9	4			
			1			2		
	9	4	2			3		
3	2				6		8	1

Sudoku # 250 - Medium

				1	6			
3	7						8	
								4
			7		1			
	1	7					4	3
	2						1	6
				5	2	3		1
	5	2		3			6	
8			9				7	5

Sudoku # 251 - Hard

		7					2	4
					1		3	
	2		5	8				
3		5	2		6	4		
			8	5		6		
6	1							
	4	2		9				5
				7		3	4	
								9

Sudoku # 252 - Hard

8	6				2			
				4	7		9	
		9	8					3
		3	4		1			
	1					5		
5		6	3				8	2
			6				7	
			7	1		3		
						8		

Sudoku # 253 - Hard

		6	8				9	
	3			1	7	5		
5							8	
		2	1		3	4		8
		4				6		3
	5						1	
6						1		4
4			2					
		7	5	3				

Sudoku # 254 - Hard

1		8	2	7		4		5
		3			4			
	6			3				7
9								4
		7	9				6	
	2	5					1	3
7				2			5	
2					8		9	
						2		

Sudoku # 255 - Hard

1		9						
8					6			3
	6							
				8		4	2	5
	2	5			3	7		
6		8		7				
						9		8
4							7	
			5	2		3		1

Sudoku # 256 - Hard

6	1		3					
	5			4				
8	2		9	6	1		5	
							1	
					5	6	8	
4	9				6			2
			1					
9		2			7			1
			5			7	3	

Sudoku # 257 - Hard

1				3		6		
				9	7			8
9			5	6	1			
3				4			8	7
	8	1				3		
		3			6			
5		2			8			9
	7		2					

Sudoku # 258 - Hard

5				3	8			
		1	6			3		7
					4		8	
	4	7	9		3			
		3	5			1		4
							3	
		2		1		6		
			4			5		
9						7		8

Sudoku # 259 - Hard

4					5	8		
		6	9	3		2		4
					6		7	
		2					5	9
1		4		6			3	
3								
							8	7
8		3						
					9	1		6

Sudoku # 260 - Hard

		4	5				7	
		2						
	8	7	4		6	1		
						6	1	
	3	5			7	2		8
			1					3
	4			5			3	
			6					9
			3					7

Sudoku # 261 - Hard

			3		8	4	1	
	8		2					5
					9	2		
7		6			1			2
				7		5		9
	5	8	9	3				
		3					6	
5		1				8		
	7	9						

Sudoku # 262 - Hard

			6			2	9	
							3	4
9		1	7			6		
	8		4	5				3
2								5
					7	8		
3			9	7	6	4		
			1		5			
1	9			2				

Sudoku # 263 - Hard

1	5			4		6		2
	6					9	8	
	8			7				
		7		8		5	4	
	1							
	7	3	2	5		8		
		5	7		9		3	
	4					2		

Sudoku # 264 - Hard

1		7					3	4
	4				1			
		8	3			1		
								7
		6	5					
4	5		6	8				
6			1		9			
						2		
7	3	1		6		9	8	

Sudoku # 265 - Hard

2					6			4
				4				
					9		2	
6	3		8					
	8			9	2	6	4	
			4				8	
3	7					4		
1						8	7	
4		5	1				9	2

Sudoku # 266 - Hard

7		9	6			3	5	
		3			2			
5							8	7
4	9			5	1			
	1						9	
3					9		4	
					5	2	1	
8				4			7	6

Sudoku # 267 - Hard

5	8							
	2			3	4			
9	4	6				3		
	3		4				2	
	1			9		5		
								4
8	9	2	3				6	
		7	8				1	
	6				5			7

Sudoku # 268 - Hard

							9	6
9	6		4	8		3		
		1				8	5	
			2					3
1								
			6		8	7		9
						6		
				1	7	9		
	3	5		4				2

Sudoku # 269 - Hard

					6		1	3
		7				2		5
8	5						4	
								4
	4		2	9	1			
	7		8		5	9		
			7	3		5		
		9					6	
		4				7		

Sudoku # 270 - Hard

1	3	5	6					
4	7				8			
					2	7		
		4					9	6
		6	3			2		4
			2					5
9			4		5			
						4		7
		3		8			5	2

Sudoku # 271 - Hard

6		5		9	7	1	2	
	7	9		4				8
	3				1		5	9
							9	
			8			6		
	9			6		5		2
	5							
			1					
		2					3	7

Sudoku # 272 - Hard

			2			3		
	9		4	8				
	5		9				2	
9					3			7
		8				2		
		1				4		
	1		8		4	9	3	
			6			1	7	
5		4				6		

Sudoku # 273 - Hard

		5		1			7	
9			2					
		2						6
1		3		8				2
	5		7	3				
						3	6	
4					1			8
	8	1			9		3	
7					8	1		9

Sudoku # 274 - Hard

9				4			7	
	1	8				4		
5	2			7				
4				9		8	5	7
							6	
1			2					
				5	2			
6	5	7	8					
		3				9	8	

Sudoku # 275 - Hard

	7		6	3		2		
	5			1				
6		3					4	
						3	8	
4		1						
	6	7			9			
3	2	8	5				1	6
9			8					
		5		2		8		

Sudoku # 276 - Hard

	5	1				9		
6		7						1
							8	
4			2					
			6				4	
				7	3	6		2
			7		4	5		
	4				8	3		
	3	8	1	5			2	

Sudoku # 277 - Hard

						9		
			8				4	
	4	1			3	2		
8		7	9			3		
	3				7	4		
4			6			1		
7		6		1			3	4
				4			5	
							1	2

Sudoku # 278 - Hard

	6			7				9
5						8	4	
4			3	9				7
	8					3		
	1				6			8
						9		
	5		6		7	1		
	3			4				
2					9	5		6

Sudoku # 279 - Hard

		8						1
	2					5		
7	5		6	1				3
			5					
		2			8	1	6	
		7	4			3		
1				8		2		9
8		9					4	
				5				

Sudoku # 280 - Hard

			5					8
		4						
1	2		7					
		9			2			3
				4		6		1
3					1	8	9	
					7			9
7				3	9		8	6
		3					1	

Sudoku # 281 - Hard

1				4			6	
	9		6	1		7		8
			7					3
					1	2		
	6						9	7
8	3				6			
					3			4
			1				5	
3			8	5				

Sudoku # 282 - Hard

6	9		7	4				
		2				8	5	
	4		2				9	
	6		1					
1				9	5	3		2
		1				6		
	2				3		7	8
8	7							1

Sudoku # 283 - Hard

	2							
			4	5				
				1	9	8		4
	3		2	9				
	9		3					
	4	5		7				3
	7	2			6			9
1						3		
	5	9	8				4	2

Sudoku # 284 - Hard

4				1		7		
		1	2				5	
	9	7						
7					6		1	
	3							5
	5			4	9	3	7	
						6		
	4		3			9		
	1			2	8	5		7

Sudoku # 285 - Hard

	2	4	3			1	7	
	6							
5						3		6
1				7				9
		9	6					
		5	9	2			8	3
			7	5	3			
	4	6	8					
						2		

Sudoku # 286 - Hard

					1			5
8				4	6		1	7
	6			2				4
3						1		
9			3		4		7	
						3		
1		8						
		2		7			9	
	4			9		8		2

Sudoku # 287 - Hard

								6
		8		3	7			
					5	1		
	9		6		4			1
		5			9		8	
2					1			9
3				9				8
						5	6	
	1	7		5			2	

Sudoku # 288 - Hard

			1		6			
7		1		9	2			8
	3			2				
9	8							
					1		9	6
	7		3		5	4	1	
2	5				9	6		7
					4			

Sudoku # 289 - Hard

			1			8		5
								7
4	6					1		
		6		8				9
	1	3				2		
				2		6		
1	8		9			4		
2	9			1	3	5		
					4			

Sudoku # 290 - Hard

7				4		5		
	5						4	
1				2	9			
5	3	1		7				
			1					
	4			5		8		
			4	9			7	2
	7				8		5	
9	6		2					

Sudoku # 291 - Hard

5				9	2		6	
			7	4				
						7	4	
				7		5		9
		2						
6			5		1	4	8	
	9		4		6			
				1		2		
	4				9		3	

Sudoku # 292 - Hard

		9					4	1
4					5			
			9					3
9	1	5		2				
	2		6		1		8	5
	6			3				
5	9	3						7
			3	7				6
6			2					

Sudoku # 293 - Hard

5						8		
				3	4		5	
		6			8	2	1	
	4		8					
	9		3		6			5
		7	9		5			
			4		3		6	8
	5	4		2			3	
						4		

Sudoku # 294 - Hard

7	3	8						2
6	5			3				1
				2	9	5		7
2	7		3		6			
		6		8		2		
4	6		2	1			7	
1					4		9	

Sudoku # 295 - Hard

			9			3		
6	1				3		2	
9		8	6					
			7		4			5
					5	2		1
4		9						8
				4		6		
	2							3
5				7			1	

Sudoku # 296 - Hard

3			9			8		
			7	3				
				6		1	5	
9				5		3	8	
4	6		2					
	1							
	8	7				5		
	3	6	8	4		7	9	
			5	7				

Sudoku # 297 - Hard

	7				3	2		
			2					1
		9	4	7	6		3	
9		2		6			1	3
7	5		8					
		3						4
			1				9	2
	6				2			

Sudoku # 298 - Hard

		7			4	9		
		3			2	1	4	
		4			9		6	2
					8	7		
	6							
1			3				9	6
			8					9
	4				6		3	
	9		1		3			

Sudoku # 299 - Hard

			1					
						7		8
	3	1	8	5				4
	5				2			
				4		8		
4	7			8	9		5	
		6					2	
			4	1		3		9
	8			3				

Sudoku # 300 - Hard

						7		
		4		3		1	8	5
9								2
3	6	7		1		9		
	9						1	
		8		5	4			
8			5	7			3	
	4				3		7	
7			8					

Sudoku # 301 - Hard

				4		9		
3	2		1		5	6		
4	6			3				
				5		3		
8			7					
		2					7	
5				6	2	1		3
	9			1	7	4	6	
1								

Sudoku # 302 - Hard

	3			5	2		4	
		8			4		3	
5						8		
7					8			5
2			3			4	7	
		6		9				3
	1			4				
	4				7			
9	6	2						

Sudoku # 303 - Hard

	9			2				5
						9	7	
7		4						8
8	2			1		7		
6				4				3
9								
			4			5	8	
	7						6	
	1				7		4	2

Sudoku # 304 - Hard

6		7		3				
					5			
5			2	1			9	
	9	8			1	2		
						7	4	
3		5			9	8		
				7			6	4
	5		1			9		8
2					8			

Sudoku # 305 - Hard

5		4						
				2	8			
		2				3		9
	6			7	5	2	9	
		1					8	5
	4	5			9			
4					1	5	7	
6					7			
				8	3		2	

Sudoku # 306 - Hard

	2	7				3	8	
	4							
8	1				6			9
				7			2	
	7	9						5
			9	4	8	6		
3								7
	8	6		2	4		9	
							6	

Sudoku # 307 - Hard

		8	9	1				
1			4					9
	9			5	2		6	
		5					1	
				3		7		4
			7			5		
8		4	3			1		
	3							
					6	8		7

Sudoku # 308 - Hard

			1	8	7	9		2
				3			1	5
9		4						
		7				8		6
			9					
				2	6			7
		3		1	8			
7		9	2					3
	2	1					5	

Sudoku # 309 - Hard

				3			4	
		9			7		8	
3		2				9		6
	1		6		3			
			1	4		8		
	2	6					9	3
							5	
8								
5			4	1	8	2		

Sudoku # 310 - Hard

		8					4	
			3	1				
	5		8					
	7					2		
3				6		1		5
5		4	7		2	6		
		9						
2	4	3	5					
				8	4	3		9

Sudoku # 311 - Hard

					1			
		5			7			1
9		3	5	6		4		
6		1						7
4	9	7		1	3	6		
		2						
5						2		
	7	4	9				8	5

Sudoku # 312 - Hard

9				8			6	
			1	3				
		2					4	1
	7	8					2	
		5		1				
	1		8		4	6		
	4	9	3					
						8		
	5		9		1			2

Sudoku # 313 - Hard

	9					6		1
		2		6		8	7	
8		1						
		8	4		9			
		5						
	2			1				5
		3				2	5	
			6				3	
5	7		2	3				9

Sudoku # 314 - Hard

7			4	9				
					7			
5		3					1	
3	4				5			2
			3	8			4	5
					9	6		
8			5			4	9	
				6				
		1	8				6	3

Sudoku # 315 - Hard

				5				2
	4						1	
2		5			3			
	7	1						9
3			6				5	
	8				7			
8		9	7					4
		3			5	7		
4	5		2		8	6		

Sudoku # 316 - Hard

2			3		1		7	
	8	5				4		
1					6		9	
					4	7		
	7			9		5		
6	3					9	1	
4					8	6		
7						3		
	9		6					2

Sudoku # 317 - Hard

9						3		5
			7		4			
	4		2			1	7	6
				1			2	9
5			8			7		1
					9	8		4
2	6			5				8
		7	9					
1								

Sudoku # 318 - Hard

6						9	3	
		4	9	6		1		2
	9							6
5					1			7
			5		8			1
4		6						9
					9			
			3	8			5	
7					2			

Sudoku # 319 - Hard

					3	7		
		1		2	6			9
3			8		5	4		
7			1		2			6
	6			8				5
			9					
	5						4	7
					1	6		
	9						2	

Sudoku # 320 - Hard

3			1					
1	9		7			8	5	
2			6					1
			4			7		
					8	2	4	6
8								3
		8						
4				3			9	
		5		9			7	

Sudoku # 321 - Hard

		4					5	
7	9						6	
	3		6					
		3	9			6	1	7
					1		2	3
4								
		6			5			
	5		8	7			9	
		7	3					5

Sudoku # 322 - Hard

2	5		3		7	6		
			1	6		7	5	
1		8						3
					5			
4	9							
7					1		8	
5		6		2				9
	3		7	5				2

Sudoku # 323 - Hard

	6			4		1	7	
	7		6		8			
5				1			9	
7					9			4
	8		1	6				
		6	4			9		8
	9					6		
				7			3	
		3	2	8				

Sudoku # 324 - Hard

			6	8				2
		9			2			
		1			5			6
		3	7		4		5	
				2				1
9			2					
2	3			4		6	8	
6	5		8					7

Sudoku # 325 - Hard

7	8							
		9			3		2	
					2			4
	4							
		8		2			7	
	7	1		8	5		4	9
				6				
	9			5	1		3	6
6	3						9	7

Sudoku # 326 - Hard

1		6					9	
	8		9		6	4	2	
9				4		7		
					8			4
2			1					
	5			2				
							8	3
		1			5		6	
5	3	9		8				

Sudoku # 327 - Hard

		7		2	5			6
		9			1			
	4	5					9	3
		4		7			6	8
7				6	2		1	
	3		1					
				5			8	
		3						7
6						2		

Sudoku # 328 - Hard

	7			4	9	2		
	2	9	3	5				
	8			7				
3			2		6		1	
		8	5				9	
			1			5		2
			7	2	3			
	4			6				3
						8		

Sudoku # 329 - Hard

	9		3				2	
	2		7			8		6
		8			4	5		7
						9	1	
8		6			5			
						6	8	
	8	5		7		1		
		9			6		5	
7					3			

Sudoku # 330 - Hard

9		2						
				5	2		1	
	7		1			3		
7		8		1	6			
	6	1	7					
	3	4					6	
	8		6		9		5	
		7	2					1
				3				

Sudoku # 331 - Hard

3	6					8	9	
							3	
			8	9	3	2		
	7		3	5				
					1			9
	4				2		6	5
		2			5			
9			2				1	
	3		7					4

Sudoku # 332 - Hard

3		8		7			2	1
		6						
2			6					4
			9				7	
		3				1	6	
	7		3		6			8
	1				9		8	
			4			2		7
7						5		

Sudoku # 333 - Hard

			2	8		3		
4				6				
	2	7	1					
3		5			6			9
7	9		4			6	5	
					5		1	
				5		2		
			6	3	2	4		
5								

Sudoku # 334 - Hard

								6
4		9				5	8	7
6	7			2				
				4	2			
5	1	4				3		
7			1	3		9		
				9				4
			7					2
8	9							

Sudoku # 335 - Hard

		1				6		
		9	4				2	
				7			5	
3							6	7
	4	7	5					
8	9					5		
						7		2
2	6				8		4	
	1				3	9		

Sudoku # 336 - Hard

	5					9	4	
					8		7	
		9	4	1	3	8		6
7								
8						7		
		6	5		2			
					4	6		
3			8		5			2
			1	3		4		

Sudoku # 337 - Hard

				1			5	
			4		3			
6	9	4		8				
			7					3
1	6				8	5		2
					4		6	
		8			2			
	4		8		5			
2					9	3		

Sudoku # 338 - Hard

4			3		8			6
					2	4	5	
			9	4				
	6	2			9			
5				1		8	3	
9								
			4	8				
	8					7		
6	4	5				1	2	

Sudoku # 339 - Hard

7			8	2				6
	2							
	3		1			5	9	
5		1		3	4	9		
						4		
			9		5			
2			5				8	
8	5		4				3	
	6				8	1		

Sudoku # 340 - Hard

								3
	2	8	1			4		5
		7		3				
			5			6	7	
				2				9
1								
4		2	8				1	
		6		9	1			
	5			6			3	2

Sudoku # 341 - Hard

5		8		4	3	7		
4	6							
9						5	8	
		4			5		6	8
			1				7	
		1		2			5	
6			5		7	9	1	
3				6				

Sudoku # 342 - Hard

2								
5	8		2	7	4			
6	1			9		7		
						6		
		4						
1					5	4	9	
	6				9		1	4
			7	6				8
	9			4	2		7	

Sudoku # 343 - Hard

		6		4	5	7		
5	9			7				
								2
7	2		9	6		3	5	
9			1				2	
	8				2			
				5				
	6	7				1	9	
					9	6	7	

Sudoku # 344 - Hard

	5		7					
3	2	8					5	
			8				6	
		7						
	1		5	3				6
2				4	8			
					6			1
4			1		9			
6	9			5				7

Sudoku # 345 - Hard

1						8		5
	8		3		4			1
	7	2					6	
	5		8		3			
	1				9			
			4				8	
2	9				5			
			6			5	2	
		6			8			

Sudoku # 346 - Hard

2				1		5		
		5					7	
	7		8		5			1
		1						6
	8	6	5	9				
7					8			9
		7						2
			2			7		8
4			7			3		

Sudoku # 347 - Hard

		3						
			1			7		8
	2			8				
7				5	6		2	
	3			9			8	7
					1		5	
6		7		2				1
			9			4		
9	5						7	

Sudoku # 348 - Hard

9		5						
	2		5				6	4
	3					8		9
					8		3	
			7	5				2
	1	3		4		5	7	
6					4	7		
	7	9					4	
				1				3

Sudoku # 349 - Hard

				1				2
2	7						9	6
		9		4		7	3	
7	6						1	
		3				6		
1	2	5						3
4			1	8			5	9
					9			
			7			1		

Sudoku # 350 - Hard

	7		1		4			
3		8	9			1		
7							3	
				5				7
2	5			3	7		9	
				6				8
	4					5	6	1
	1				5			3

Sudoku # 351 - Hard

		9				3		4
						7		
2		4						6
6			8	7		4		
	1							2
		3		2				5
	8							3
	2		4	8		6	1	
	9			5			4	

Sudoku # 352 - Hard

		8	9	1				7
		1	6					5
	2	5	4			9		
					1			9
	8	9				7		
5						8		
1								
	7						6	
3			5	8				2

Sudoku # 353 - Hard

5	3		7					
								8
			5	3	2		6	
			9	6				
3								
6	1		8	4				7
		7				9		6
1			4	7				
2	8					1		

Sudoku # 354 - Hard

			7				8	
		6				9		
5			1		3		2	
			9	3		5		
7	4				2			9
					4			6
				4			3	5
1	2					4		
						8	6	

Sudoku # 355 - Hard

	7		3					4
	2		1			8	5	3
3			2					
7							8	
				8		9	3	
9	1							
8				6	5			
						3		5
		5			7		9	

Sudoku # 356 - Hard

		3			5			
	8			9				5
6				2			9	
		6						4
			1			5	6	9
	7		5				8	
	5		7					
			8		6	9		
8						1	7	

Sudoku # 357 - Hard

	6						2	
			7		3			
	9		1					6
		2	8			1		7
				7	9	8		
		5	4					
3						7	9	
			3	2			4	
	5		9			2		

Sudoku # 358 - Hard

5	4						7	
		8		2				
			1			4	3	
					7	3	8	2
			8					
				1	9	7		
	2	3			1		6	
	1		7		3	9		
7	9							

Sudoku # 359 - Hard

9			3	2		6		7
	3		7			1	2	
		1		6				
								5
5	1						4	
	4			7		9		2
1			5				3	
				4				
			2		3	5	6	

Sudoku # 360 - Hard

		9	7		6	2		
				5	2			
			1			6		
8						5		3
4		7					2	
9							1	
		8	3					4
	7						9	
5				1	9			8

Sudoku # 361 - Hard

	7	3		2	4			
	9					8		
			8				4	
	3				1	9		
	1	2					3	
		5		8		6		
			4					
	5	1	2			4		
			1		7		6	

Sudoku # 362 - Hard

	2							
7		8						
	9						3	
		2		6	1	3		4
	7		3		9	2		
		4	2	8				5
	3		4			1		
	6			7		4		3
						9		6

Sudoku # 363 - Hard

3			2			1		6
			1				5	
		4	5		8			
8				9				
4	9					3	2	
				1				
6					1	7		9
	8							3
		3					8	2

Sudoku # 364 - Hard

		2	9					
						3	9	
	6				4			5
				2	9			7
	1	7		8			2	
				7		9	5	
			3					
8		6	2	1				9
5			8				1	

Sudoku # 365 - Hard

2			8		6			
	9			4		6		1
			3					7
8		3		9	4	5		
	1						3	
9	8						7	
	6	4		7			9	
			6			8		3

Sudoku # 366 - Hard

	7			6	1			
1	4		2		3		7	
2		9			8		6	
4		3						
	2			5				7
9								
				1				
			7				2	8
7	8						3	9

Sudoku # 367 - Hard

		6	7			3		
1		2	5		3		9	6
	9					8		
	2	7			6			
	5						7	
		1				6	5	
			8		1			
		5						
3	1	4				7		9

Sudoku # 368 - Hard

9	5		7	3				
	1	3			6	7	5	2
7			6		5		3	
						6		
8		6		9	1	5		
					8	2		
				1				
		9	3	6		1		

Sudoku # 369 - Hard

				3	5			
2		1						
9			6				4	
4			7				6	3
			2	9		7		
	5					9		
					9			
	4			6				2
	9		3		4		8	6

Sudoku # 370 - Hard

9			3				7	
					4	8	3	6
					2			
		1		9				
4					3	5		7
3						2		
		5					8	
	6	9		4				
	8					7	5	4

Sudoku # 371 - Hard

		1	4		8			
				6	3	8		
4				9				7
1	7		5					
5	3		6			4		
8				3	1			
	5		9		4		7	
								4
		7					3	

Sudoku # 372 - Hard

				3	9			
								7
8	9	6		1				
							4	3
	5		2	4		8		
			8		3		7	
	7	1					8	
5	6				7			
		9			1		3	6

Sudoku # 373 - Hard

							4	
	3		5			8		
			7	8		2	6	
	6				7	9		
3	7	8	9					
	2					1	5	
1					8			
7	5			4				
	4			9				

Sudoku # 374 - Hard

		1	7	8				3
						5		
7				3	6			
1	3		2				5	
		6				7	2	
							8	
4		9			3			
				5		6		
	5			4	1	9		

Sudoku # 375 - Hard

		4		1				
		9	6		2			7
2	7						1	6
				5		9	6	
		1	3			8		
			9				4	
	9	2			8			
1		8						
3			2		9		8	

Sudoku # 376 - Hard

7	8			3				
2		5		1			4	
					2	3		6
		3				1		
9				6				8
	7				3			
		2	1				9	
	6		4	9		5	8	

Sudoku # 377 - Hard

	3	7	4				8	
		9	5	3				
		2			7			
		1				7		2
	6		2				1	5
					6			8
		6		4		3		9
	2	4			8			
							5	

Sudoku # 378 - Hard

			3	7		5		
	6		9				4	
1	2			4			3	
					4			2
6			5				7	
3						9		
				3			1	8
	1				7	4		
2					6			

Sudoku # 379 - Hard

2						9	7	
	3		9			5		
9			6					
	6	3	4	2		8		
1		8						6
	4		7			1		
	5						9	
			1	8		2		
	7	2			6			

Sudoku # 380 - Hard

		5			6			9
				7	5	4		6
		6						
9			6					
1				8		2		
6	8		3		1			5
			8		4	6		
	1						3	
			7			1	5	

Sudoku # 381 - Hard

2		9						5
		7	5				6	2
	4		3					7
6	9	3	1		4			
1			8					
	7					4		1
	8		9	5		7		
					8	6		

Sudoku # 382 - Hard

				2			8	
9		3	8			6		
4			9		6			1
1								3
	4	7					9	2
	9		1				7	
				5		2		7
			6		9	5	4	
			3					

Sudoku # 383 - Hard

		5		2				
							1	2
2	3		7	9			8	
					4			
			5	8		9		
4	2			3				
3	1							6
		7	6		9			
5			8			7		

Sudoku # 384 - Hard

	9	4				6		8
	3				5			
				9			5	2
	1							
		5	7		3	1		
6	4				1		8	
		1					7	
			4					9
9	2					5		

Sudoku # 385 - Hard

		6					9	
	2			1		6		3
			8	3		4		
				4				
9					2	5	8	
								6
4				2	3			9
7		1				3		
			5		4			1

Sudoku # 386 - Hard

				7	1			
	5	1			9			
	4				7			2
7			8			1		5
		6					8	
9	7				4	3		
8				3		2		9
4			5					1

Sudoku # 387 - Hard

		9			4		6	1
2								
	8		9		3			
3		7	8					6
8			2		7	4		5
				5				9
	2		5				9	
		5				1		4

Sudoku # 388 - Hard

		6		4				1
		4	8		5			7
9	2	7		6				4
		9	2		1	7		
7							8	
				9				
1		8				6	7	
6							4	
					4			8

Sudoku # 389 - Hard

2								
	3	9		4	1			
	5			9			3	
		2			8		4	9
			5	1				7
7					2	3		
						4	9	
			8					5
4		1		5		8		6

Sudoku # 390 - Hard

		1				8		
			2			6		
3			4	1				
			9				5	
	5			2		4		
		6	3	5		2		8
		7			6		9	
						1		
6	9	5	7					

Sudoku # 391 - Hard

			1	2	8			
								4
6		9	5	7			8	
	3			1	2	7		9
					6	2		
	7	6		3				
5	6							
	9					8		5
						1	2	

Sudoku # 392 - Hard

		4						
		1	3		7	4		
	9						8	
	1		4	6		5		
4				9	8	7		
	2						4	
7		6			9		2	
	3				6			
	8				3			

Sudoku # 393 - Hard

		8		5	4			
3	6			7				
		1						
			4	2		9	1	
4				1			5	
8					3			6
	2	7	1	3				
1							6	
		4				7		

Sudoku # 394 - Hard

1		6			5			
2						4		
5					4	9		
	1	2			8	7		
7	9				1			
8						5		
4					3		6	
				2	6			
	6			1				2

Sudoku # 395 - Hard

8	5							4
					1			
				7		6		
		6	4	1				
2	9	4						8
5						4		3
	4				3		8	
		5		2	4	7		
			1	8		3	4	

Sudoku # 396 - Hard

2			1			4		
	4	9		3	8		1	
	7						2	3
	1							
			6		2	1	8	9
						7	6	4
					1		9	
	2		9	7				5
	6							

Sudoku # 397 - Hard

			5		2			6
								3
	2						4	8
7								
5	1					7	8	
					8			9
4			1		3	8	6	
	3		4		7		1	2
				9	6			7

Sudoku # 398 - Hard

					8			
1			5				9	
2	5			3				
3						1		
		7	4			5		
			1	9			4	8
				2	6			
	9	2			1			
4	8				9		1	

Sudoku # 399 - Hard

		5				6		
1			7					
9			3	2	1			
		8	2					7
	9	1		6	8		2	
	4							
			1	4				
6				9		1		3
					3	9		

Sudoku # 400 - Hard

	6	5	1	2				
							3	
		3	5		4	2		8
			4		7		6	
3					6	7		9
		7				3	2	
				1	8		9	2
2		6						
				4				

Sudoku # 401 - Hard

	3	4		2				
			8			4		
	7							
			9	1			5	
		8	7	4		3		
		5					2	
	8		2		9			3
9		6				8		2
7					6			

Sudoku # 402 - Hard

1			2				7	
	5			1		4		
	7			5	9	3		
								8
						9		7
	1	3	7		5			
					6	8	2	
		9	1	3			6	
		1				5		

Sudoku # 403 - Hard

				5	7		3	
								8
			1	9		2		6
2				1		4		
	8				5		2	
	9	6	4			3	8	
		7	5				4	
	4				3		5	
9		2						

Sudoku # 404 - Hard

4		8				7		
	7			5				3
2	9							
				3				4
9			1		4			
		1	9					
							2	
5	3		6			9		
6	1		2				4	8

Sudoku # 405 - Hard

				9	3			
				2		7		
3	6		5		7	9		1
							8	3
	3	9		5	2			
	4							6
	5					8	1	
2		8						
	7			6	8		2	

Sudoku # 406 - Hard

								3
6								5
			9		3	7		2
3			8		2			
		7		9		8		
				7			6	4
9		5			7			
	3					2		
2					8		4	9

Sudoku # 407 - Hard

				9	3			
7								
				2	4	3		
	5		4				3	
2	9		8					
		4		5	9			
	7					4	6	
		9		3	1	7		
		1	6			9		

Sudoku # 408 - Hard

7		6						
	2	9	1			4	5	
8			9				6	
6	1						8	
				1		3	9	4
		4			2	1		
2			7	3				
			5		6			1
						9		

Sudoku # 409 - Hard

8								
		1	3	2				
			1	6	9	8	2	
9	3		5					4
4	6							
2		8					6	
5	4			1	2	7		
	8			5				
		2	4					

Sudoku # 410 - Hard

			1	9	7	8		
6	1		4					
	3				2			
		9			6	1		5
	5		9	3				
3					1		9	7
4								9
							7	
2			3				8	

Sudoku # 411 - Hard

9			6			8		7
			5	4	9			2
	3		7				4	
		4	9				1	5
7				2				9
						6		
	1	6						
		9						4
	5	3	2				7	

Sudoku # 412 - Hard

2		9			3			4
	5		9					7
			2				9	
4	6					7	1	
			3		1			2
1			4					
	4					6		8
8			1		5		4	
						2		

Sudoku # 413 - Hard

		3			4			2
	1		9		5			
	5	8					1	4
5	8			3				
	4	6	1		2			
8								5
9				7	3			
	3	1	5			2	4	

Sudoku # 414 - Hard

	6				7		1	
					2			3
7		1	5			6		4
				1	8			
3								
9		4				8	5	1
5								
		9		2		5	4	6
						2	9	

Sudoku # 415 - Hard

	2					6		
6	1						8	5
5			8	1				9
8		6		4				
		3	9	8			7	
1				6	2	5		7
	9	2	4	5		1		

Sudoku # 416 - Hard

	1			4				6
3		8	6				1	9
		4				7		
	4	5	9					
			4			8		
	3	7						
			7		9	3	6	2
	5				2			4
		6						

Sudoku # 417 - Hard

				2	3		8	7
		5		1				
3		6				2	1	
		1						
			7					1
					8	4		2
				5				3
		3	2			9		
	9	7	4		1	5		

Sudoku # 418 - Hard

		7						
	5	2		8		1		
3			1	9			5	
	2			5	6	9		
	1		9		3		2	6
	3				8	5	7	
		6	3		5			
2							3	

Sudoku # 419 - Hard

8					6			
			8	5				
	2	5	1			7	3	
3			9				1	
		8				4	5	
		9			4	3		
9		7				5		
					3	8		2
	8		6		7			

Sudoku # 420 - Hard

7		8	4			6		2
				5				
	3							1
8				2		1		
5	4				8			
3	6						9	
9				7		4		
					5	7	8	
			6			9		

Sudoku # 421 - Hard

1				4				6
5		2		1		4	3	
	4					9	5	
7					8		9	
				6	2			5
		6		9				
	8		1	2			4	
	5				3		1	

Sudoku # 422 - Hard

8	9			3	6			7
						3		
						2		
	6	1					9	4
5			6	9			7	
				2				5
	8				7			3
7				1	4			
4		5	3				2	

Sudoku # 423 - Hard

2	1						9	
6	9			2	5			7
8		3	6					
	6		9		8		3	2
		8			7			
				1				
				6		4		
9								1
	5			9				

Sudoku # 424 - Hard

		6					2	4
8				4	7	6		
	4			3				
7								6
3						1		
			7	2	4	8		5
	7	3						
	8		9					7
	1		5		6		4	

Sudoku # 425 - Hard

9	1			7	6			
6	5	8	1					
	8				4		5	
		5	3					7
4			7					
		4					8	
			8			3	4	
	9	6	4	5				

Sudoku # 426 - Hard

4		7		5	8			
					3			7
							8	
5						8	6	
	1	2			4			
		3				4		2
	2			9				1
			1	3			5	
		6		2	5			

Sudoku # 427 - Hard

		3	2					8
				3		4		7
			5					
				5		6		
5			8		1			
7					2			5
1	5	6		2				
		8	6					3
4				1		9		

Sudoku # 428 - Hard

8		7	1					
					3	8		4
	2						1	
6						5		1
	1	8			2		4	
2								
	5		4	8				
					9		8	5
		3				6		9

Sudoku # 429 - Hard

3				6	7		1	
				3				8
		9				4	6	
4	1			9		3		5
				7				
		6	1					
		5	7		8			1
					1			
1		3			6	5	4	

Sudoku # 430 - Hard

	3				1			
6			8					9
9			7					
	5					8		4
		3	6					5
	8		5		4	7		3
					8			
	4		9	7	5			
1		7			6	2		

Sudoku # 431 - Hard

		2			7			6
	8			6		3		2
			4					
	5	6					3	9
7	9				1	2		4
	7		5	1		9		
	6	4	3	7				
				2	6			

Sudoku # 432 - Hard

				5		6		
9					1			
	5	1				8		9
	9					4		
		6	3	2			9	8
3				6				
				1				6
				9	5			
	3				8	9	5	

Sudoku # 433 - Hard

							5	
6	3		4		5		8	
	4			8	7			1
						2		6
		7	5				3	
	6	9		3				
	8		1					5
			3			7		
		2	7			1		

Sudoku # 434 - Hard

							9	8
4			2		9		7	
						6		3
	7		1					
	3			4		7		
				8	5			1
1			9					
		4		7		9	8	
				5	4			6

Sudoku # 435 - Hard

5								9
		6					7	
		3			9		6	1
4				9	2		1	7
				5				
				1		6	3	8
	8			6				
		1		2				4
	4				8	1	9	

Sudoku # 436 - Hard

8		1		9			4	6
4	5							
9		4	7			6		
	7			4		3	2	
	2				6			
					3			
2		5			8			4
					4	8	5	

Sudoku # 437 - Hard

9					7			
			3	1				
						1	7	
5		7	4				8	3
					1			4
			7				1	
	8		9				3	
	2				4	5		
	4				2	8		7

Sudoku # 438 - Hard

9						2		
7			3					9
			1		5	7		
			9	1				4
		8		6	2			1
1		6	8		3			
							4	
		5		8	1			2
	1		4			5		

Sudoku # 439 - Hard

	4			9				
					6			5
			1			8	6	
							8	6
1			4		7		5	
6	2		8	3				
			2					
	7			5		6	3	
		2		4	8		1	

Sudoku # 440 - Hard

		5	2		1			9
4		2				8	1	
			7	8				
		6			4		8	5
			1					7
			9					
		7	4					
3					9	6		8
2							3	

Sudoku # 441 - Hard

		5			1	7		
	7	4	2					
6				5	8		2	
						1	7	
		7				9		
				7				8
		8	4				1	
5		3						
2		9	3		7	4		

Sudoku # 442 - Hard

		2						
		1			8			4
6	5	8				9	2	
			1		7			
2			6	8				
				3		1	6	7
	9			7			4	
		6		4		8		
	2		9			3		

Sudoku # 443 - Hard

6		8		9	4		1	
		3	8				5	
	9			1				8
						2		
1	8		3					
		4			9			
8	3	1			6		9	
	5	2					4	
					5	7		

Sudoku # 444 - Hard

1		7	5		9			2
					6			1
					1		9	
	4							
7				8		9	4	
		1				7	2	6
					5	4		
	6			1				
4			2			8		9

Sudoku # 445 - Hard

9	8				3		2	
	3			8				4
				7		6		
2	9	1					4	3
4					5	8		
		2	6					
5					7	2		9
	1	9		5				

Sudoku # 446 - Hard

		8	4		1			
						7		
				7				1
5							1	
	3		6	1	4			
	8		7		5	6		3
6	5	9					3	
							6	
		7	8	4		5		

Sudoku # 447 - Hard

6		3			1			4
	7				9		5	
							2	7
1		9	7	5			8	
4		5						1
			1		8			
3								6
			2	1	5			
			9					

Sudoku # 448 - Hard

8			4	9				
				6			8	
1	6	7		5				9
7						1		
4	8	5			7			
6		2						4
				2		5	6	
					8			3
							2	

Sudoku # 449 - Hard

1		7		4		5	2	
3		6						
	4		9				7	
7	5		8		6			
6			3		2	7		1
					7	2	8	
		2			4	9		
	1							

Sudoku # 450 - Hard

				5				6
1	2		9					4
	4	6		2	8			3
					5			8
3					6	1	7	
		4						
					2			
6					9			2
2	7		5	4			8	

Sudoku # 451 - Hard

			2	7	6			
3							9	
	5	6						
		3			8			
			6		2			1
		7		3	9		4	6
2		9					3	
		5			3		1	9
8					4			

Sudoku # 452 - Hard

3						5	2	8
				5		3		
	5			4			9	1
				9	8			
			5					
		7	1				6	
5	4	2	6				3	
		1	3					9
	3			7			5	

Sudoku # 453 - Hard

	6						9	
8					3		7	
	4		8					
			6		7			1
	9						3	
		8		1	5			
3		2	4				1	6
					2			
4	5	1	9				2	

Sudoku # 454 - Hard

1								
				9		7		
4					1			8
	9		2	6				1
2	8					6		9
		7				4	8	2
		5	8		2		9	
9			6					
	3		4					7

Sudoku # 455 - Hard

7	9	8					1	2
2			7	8				
5			9		6			7
			5				6	
		9			2			4
	4		6					
						2		
1		2		5	3			
	3				9			5

Sudoku # 456 - Hard

		2	7					8
7		8		2			6	
3	1		6				2	9
							9	5
			1					
	3				2		4	
	9			1	5			
2		3			9		8	7

Sudoku # 457 - Hard

			1					
8					4		7	
	2			9	3		5	
				4				
2				7		6	1	
			8	5	2	4		7
	5	2	4					
		8					2	
	7	3					8	1

Sudoku # 458 - Hard

	1					3	9	
					6			
		3		4			2	
5		1			7			6
								9
	9	4				1		
	8			5				
	4		6	3				8
		5	8		4		7	2

Sudoku # 459 - Hard

		5		9	6			7
	6		5	7				
	7			2			9	
6		2	8			1	5	
1								
							7	
		9	7				2	5
2				1				
		3		6	2	4		

Sudoku # 460 - Hard

6			7		4			
		4		3		6		
8							3	
3					8			
		2	3					6
	5		4		7		9	8
5	1				2	9		
					5		7	3
			6			2		

Sudoku # 461 - Hard

	5		9		3	1		
			5		7	3	4	
	8					7	2	
	9					6		2
							3	
3		1		4				
1							7	
		2		3		5		
			1		2		8	4

Sudoku # 462 - Hard

4			1	7		9		
					3	6		
3						4		
		7		6	8		9	
	3	2			7			
				2	5			
		5	7				1	
2				9			8	
9								3

Sudoku # 463 - Hard

				2				
				4	9			
	4	9	6	7				1
	1					3	7	
	8		4					9
9					2	1	5	
3					5	4	2	
		8					9	
	5				6			

Sudoku # 464 - Hard

						1		
	3	4				9		
				9	7		6	
			7			8		1
7								6
		2		3			9	
2			6					
	5				3		8	
6	8		2	4	9			5

Sudoku # 465 - Hard

8			1			5		
		7		5	3	1	2	
				9		8		
7							3	
				6				1
3			7	2			5	
1	5		6					
6						4		5
		2			4	7		

Sudoku # 466 - Hard

						1		
4								8
		2	5	1			6	3
							5	
2		7			3			6
			4	6				7
		6		7		8		
	7					6	9	
3	5	4						

Sudoku # 467 - Hard

		8		1	7			2
			8				3	
			2		3	7		8
	9			7			1	3
		6						
7						5		6
	2	4						
	5			9				7
3			4	8			6	

Sudoku # 468 - Hard

7			9					1
			5	2			3	
6			1					9
	7	2						6
			8	3	5	2		
5		4					9	
					1	8		3
		8		4				
		9					7	

Sudoku # 469 - Hard

8			3	1	2			
							7	
						1	3	
	8			2		6	9	
1		5	8					7
6		7						
					7	8		5
		8	1	4				9
		4	2		8			

Sudoku # 470 - Hard

	3						4	
		4	6			5		9
				8				7
	2			4		1		6
					2			
9	1			5		3		
	6		3				8	
		7	2			4	1	
	5					6	7	

Sudoku # 471 - Hard

	1							
					4	7	1	8
		4			5			
	7	9		3	2			
			5				8	
	5				6	4		
3	9						5	
2		5	9	7	1			
						6		

Sudoku # 472 - Hard

2			5					6
	5	8						
					7		5	
		2				4	3	
				8	9			
	1	6			3		8	2
4	2		6			1		
			8					4
			4		1	9		

Sudoku # 473 - Hard

				8		3		
2	8				3			
			2		6			9
6				1	9			
	3		6		4	1		
								5
	6	4		5		7	1	8
	9	1		3			4	

Sudoku # 474 - Hard

			7	8	9			
5							9	
1		6						
		1			4	2	3	
9					1		8	7
			2			5		
			6		7			
6	8						4	
	2		1					3

Sudoku # 475 - Hard

	9	1					2	
7								
			3		9			6
		5				1		
	3				1			9
		2	6				3	4
	6	3	8		5		4	
5				6		2		
		9		2				

Sudoku # 476 - Hard

	9	5		4		3		
1					6			
			3	7				
7	6			1				4
8				6				9
3	5				7			
		6		2		1		5
2	1					8		
						4		

Sudoku # 477 - Hard

6		1		5		3	9	
8								7
						5		
1		9	4	8	2			
					5			
3		7	1					4
	9		3		4			
	3					6		
			5			2		

Sudoku # 478 - Hard

8	6				2			
2		3	8		5	9		
6	7	9	4	1				5
3							4	
5			7		9	6		
				7				6
						3		
		2				5	7	1

Sudoku # 479 - Hard

			3					
7					8	5	2	
	5				2	6		
4					7	8		
		1		8			4	7
		8		6			5	
3		6		5				
	4						1	2
				4				

Sudoku # 480 - Hard

	6		7		2			9
		8					7	
7	2					6	3	
					7	9		
	5				8			
		6	2	9		1		7
			1	8				
4					6		8	
	9					5		

Sudoku # 481 - Hard

3				6				
		8				5	2	
		2	3					7
4					1		9	
	3			9			5	
2	9	5			3			4
				5	9		3	
	2		1					
					8	4	6	

Sudoku # 482 - Hard

6							4	7
5			8		4	1		
4	7		5	9		3		
	5		3				6	
		9	4					
3	4							
2					3			
					8			1
8			7	4			2	

Sudoku # 483 - Hard

	2	3	7	6				
			2					
7	5			8		6		4
	9					4	6	
	7			9		5		1
		6			2	3		
							4	
6	4							7
		8			1		5	

Sudoku # 484 - Hard

				1		7		8
	7	5			3			
		2	9			4		
	2		6				4	
	9					6		
	1		7		5	9		
		1		2				
			5				3	
					8	2	1	6

Sudoku # 485 - Hard

9					1		8	6
8				3			4	9
2				5				
				8				3
		1			4		2	7
						4	9	
1	7				9			
		4						
			6	1		7		

Sudoku # 486 - Hard

			2					
	8	5		1			3	
	1			4	7			
8		4						2
	2		3			9		7
		9			8			
				9		5	2	
9	3							8
						4		3

Sudoku # 487 - Hard

4			7	6				2
9				4				
6		1					8	
				1			3	6
	1				4			7
5					3			
						7		
	4	6					1	8
		8		7	9			5

Sudoku # 488 - Hard

			2		6			
	2		3			7	6	
		8	1	9		4	3	
5				3	1			7
	7	9					2	3
6						8		
		6		1	2			
			5				7	
8								

Sudoku # 489 - Hard

	7	8				2		
					9		3	
1		2						
			7	9	8			
8		7					6	
	5					1		
	8	5			2		7	4
				6		3		
		1	4		7		5	

Sudoku # 490 - Hard

					9		1	
			6	1	4	3		
7			8					6
		8			7			
	9	7					6	1
3								4
	3	2	1		8	6		
						9	3	
				3			5	8

Sudoku # 491 - Hard

					2			
	9	8			3	1		
3				6				
5					8	4		7
			6			9		3
			9		4			
	6		2	8			7	5
				7	9			
	1	5						9

Sudoku # 492 - Hard

	8		1				5	6
		7	6			9		
	1							
7			3	6		2		
				7	4			
					2	3	9	
			7	9				
							4	9
	6	3		2		1		

Sudoku # 493 - Hard

							5	
9			2			6		
4		6	8					2
	3			4	8			
2					3	4	8	
6					1			
						7	1	
1							4	
		4	7			2		8

Sudoku # 494 - Hard

2		7						
				4			3	
		3	1	8	9	4		
4	2		7			1		
1						3		
		8				5	2	
		1				9		6
5	8		9				4	
			8					

Sudoku # 495 - Hard

2	4			8				7
7		5	4		1			8
					7		3	
8	2	9						
			5	4			2	9
	5					7		
6	8							2
		2	7	3				6

Sudoku # 496 - Hard

7					3		8	4
	3	8			1	6	5	
6	4	9		2				
			5				3	
1			3		4	7		
			2					
4	7	3		6		1		
		1						6

Sudoku # 497 - Hard

8		6		3		9		
	4			2				1
				7			6	3
3	9				2			8
2	5			1			7	9
					6	3		
		9						6
5		1					8	

Sudoku # 498 - Hard

1	5		2	9		3	6	
	2		1					
					3			9
						7		
3				2	9		8	
	6	7			4			
	3	1				6		
							2	
6			9		2			8

Sudoku # 499 - Hard

		1					4	
				9			1	6
			3		7	8		
9	3						7	
				7				
2	8						3	5
4		2			1			
	7	6			8		2	
		3		2	6	7		

Sudoku # 500 - Hard

		2			8		5	1
8				2	5			
5		7	9			2		
				5		9		
3				1				8
	2				7			
	4		1			7		
	9						2	
7						3		

Sudoku # 1

5	9	8	4	1	6	2	3	7
3	2	4	7	8	5	9	1	6
6	1	7	3	9	2	5	4	8
8	5	6	9	7	1	3	2	4
7	3	2	5	6	4	1	8	9
9	4	1	8	2	3	6	7	5
1	7	9	2	5	8	4	6	3
4	6	5	1	3	7	8	9	2
2	8	3	6	4	9	7	5	1

Sudoku # 2

8	4	7	5	2	3	9	6	1
5	1	6	7	4	9	8	2	3
2	3	9	8	6	1	7	4	5
6	9	3	4	1	2	5	7	8
1	2	5	3	7	8	4	9	6
7	8	4	9	5	6	1	3	2
4	6	8	1	3	7	2	5	9
3	7	1	2	9	5	6	8	4
9	5	2	6	8	4	3	1	7

Sudoku # 3

6	7	2	3	5	1	9	4	8
3	1	8	6	9	4	7	2	5
5	9	4	2	7	8	1	6	3
2	4	1	7	3	5	8	9	6
7	5	6	8	1	9	4	3	2
9	8	3	4	6	2	5	7	1
8	2	9	1	4	3	6	5	7
4	3	7	5	8	6	2	1	9
1	6	5	9	2	7	3	8	4

Sudoku # 4

5	7	4	9	3	1	6	2	8
9	2	1	6	4	8	5	7	3
8	3	6	2	5	7	4	1	9
4	1	7	8	2	9	3	6	5
2	9	3	4	6	5	1	8	7
6	5	8	1	7	3	2	9	4
7	8	2	5	1	4	9	3	6
3	6	5	7	9	2	8	4	1
1	4	9	3	8	6	7	5	2

Sudoku # 5

3	6	4	8	2	9	5	7	1
8	7	5	6	3	1	9	2	4
9	2	1	4	7	5	6	3	8
5	3	6	2	9	8	4	1	7
1	9	8	7	4	3	2	6	5
7	4	2	1	5	6	3	8	9
4	5	7	3	1	2	8	9	6
2	8	9	5	6	7	1	4	3
6	1	3	9	8	4	7	5	2

Sudoku # 6

3	1	4	2	7	5	9	6	8
8	2	9	6	3	1	7	4	5
7	6	5	4	9	8	3	1	2
1	4	6	3	2	7	8	5	9
5	7	2	9	8	6	4	3	1
9	3	8	1	5	4	6	2	7
2	8	1	7	6	3	5	9	4
6	9	7	5	4	2	1	8	3
4	5	3	8	1	9	2	7	6

Sudoku # 7

7	4	2	3	9	8	1	5	6
8	3	5	1	7	6	4	2	9
1	9	6	5	4	2	7	8	3
6	2	7	4	5	3	8	9	1
4	8	1	9	6	7	5	3	2
3	5	9	2	8	1	6	4	7
9	1	8	6	2	5	3	7	4
5	6	4	7	3	9	2	1	8
2	7	3	8	1	4	9	6	5

Sudoku # 8

5	1	7	3	6	8	4	9	2
6	2	9	4	7	5	8	3	1
8	4	3	2	9	1	6	7	5
9	3	5	6	1	2	7	4	8
4	6	8	7	5	3	2	1	9
1	7	2	9	8	4	5	6	3
7	9	1	8	2	6	3	5	4
3	8	6	5	4	9	1	2	7
2	5	4	1	3	7	9	8	6

Sudoku # 9

7	3	6	9	1	2	5	8	4
5	9	8	7	4	6	1	3	2
2	1	4	8	3	5	9	7	6
3	8	2	4	6	1	7	5	9
6	4	9	3	5	7	8	2	1
1	5	7	2	9	8	4	6	3
4	2	5	6	8	9	3	1	7
9	7	1	5	2	3	6	4	8
8	6	3	1	7	4	2	9	5

Sudoku # 10

6	1	8	3	9	2	7	5	4
9	2	3	7	4	5	8	6	1
7	4	5	6	8	1	3	9	2
1	5	9	8	3	7	2	4	6
8	3	4	1	2	6	9	7	5
2	6	7	9	5	4	1	3	8
3	9	2	5	6	8	4	1	7
5	8	1	4	7	9	6	2	3
4	7	6	2	1	3	5	8	9

Sudoku # 11

2	3	8	1	9	4	7	5	6
7	1	9	5	2	6	3	8	4
4	6	5	3	8	7	2	9	1
5	4	2	8	7	9	1	6	3
9	7	6	2	3	1	5	4	8
1	8	3	4	6	5	9	2	7
6	5	7	9	1	8	4	3	2
8	2	4	7	5	3	6	1	9
3	9	1	6	4	2	8	7	5

Sudoku # 12

1	3	6	8	4	2	9	5	7
8	9	5	7	6	3	4	1	2
2	7	4	1	5	9	3	8	6
5	6	1	4	7	8	2	9	3
9	8	2	3	1	6	7	4	5
7	4	3	9	2	5	1	6	8
3	1	8	5	9	7	6	2	4
6	5	9	2	3	4	8	7	1
4	2	7	6	8	1	5	3	9

Sudoku # 13

7	2	6	9	1	3	5	8	4
5	9	4	8	2	6	1	7	3
8	3	1	5	7	4	9	2	6
1	4	5	2	9	7	3	6	8
2	8	7	6	3	5	4	9	1
3	6	9	1	4	8	2	5	7
6	7	3	4	5	2	8	1	9
9	5	8	3	6	1	7	4	2
4	1	2	7	8	9	6	3	5

Sudoku # 14

8	9	1	5	4	6	7	3	2
7	4	5	9	2	3	1	6	8
6	3	2	8	7	1	4	9	5
9	6	4	7	3	8	2	5	1
2	1	8	6	5	4	3	7	9
5	7	3	1	9	2	8	4	6
4	5	9	2	1	7	6	8	3
1	8	7	3	6	5	9	2	4
3	2	6	4	8	9	5	1	7

Sudoku # 15

6	4	2	9	8	1	7	5	3
8	3	7	4	5	6	9	1	2
9	5	1	7	3	2	8	4	6
2	1	6	8	7	3	4	9	5
3	9	8	5	6	4	1	2	7
5	7	4	1	2	9	6	3	8
4	2	3	6	1	7	5	8	9
7	8	9	2	4	5	3	6	1
1	6	5	3	9	8	2	7	4

Sudoku # 16

3	4	2	7	9	1	5	6	8
8	5	1	6	3	4	9	7	2
9	7	6	8	2	5	4	3	1
6	3	7	4	8	2	1	5	9
4	1	9	5	6	3	8	2	7
5	2	8	1	7	9	3	4	6
7	8	3	9	4	6	2	1	5
2	9	5	3	1	7	6	8	4
1	6	4	2	5	8	7	9	3

Sudoku # 17

8	3	9	6	2	7	4	1	5
4	1	5	8	9	3	6	7	2
7	2	6	5	4	1	9	8	3
1	6	2	9	7	4	3	5	8
3	5	8	1	6	2	7	4	9
9	4	7	3	5	8	1	2	6
5	8	4	7	3	9	2	6	1
6	7	3	2	1	5	8	9	4
2	9	1	4	8	6	5	3	7

Sudoku # 18

9	7	4	2	8	1	6	5	3
5	1	2	7	3	6	4	9	8
8	3	6	5	9	4	1	7	2
2	6	5	8	1	7	3	4	9
1	8	3	9	4	2	5	6	7
7	4	9	3	6	5	8	2	1
4	5	7	1	2	8	9	3	6
6	9	8	4	7	3	2	1	5
3	2	1	6	5	9	7	8	4

Sudoku # 19

6	5	1	3	8	4	2	9	7
3	4	2	5	7	9	8	1	6
9	7	8	2	6	1	5	3	4
2	3	5	4	9	6	7	8	1
1	6	7	8	3	5	4	2	9
4	8	9	1	2	7	6	5	3
5	1	6	9	4	8	3	7	2
7	9	3	6	5	2	1	4	8
8	2	4	7	1	3	9	6	5

Sudoku # 20

7	2	5	3	8	9	6	4	1
6	1	3	2	4	5	8	9	7
8	4	9	1	7	6	2	3	5
5	3	6	4	2	8	1	7	9
4	8	2	9	1	7	5	6	3
1	9	7	5	6	3	4	8	2
9	5	8	6	3	2	7	1	4
2	6	1	7	9	4	3	5	8
3	7	4	8	5	1	9	2	6

Sudoku # 21

1	6	2	4	5	9	3	8	7
9	8	5	2	3	7	6	1	4
4	7	3	1	6	8	5	2	9
6	3	7	5	4	1	8	9	2
5	9	4	8	7	2	1	6	3
8	2	1	3	9	6	7	4	5
7	1	9	6	2	5	4	3	8
2	4	8	7	1	3	9	5	6
3	5	6	9	8	4	2	7	1

Sudoku # 22

3	7	8	9	4	2	5	1	6
6	4	2	1	3	5	8	9	7
5	1	9	8	7	6	4	2	3
2	9	7	4	6	8	3	5	1
4	3	1	7	5	9	2	6	8
8	6	5	3	2	1	7	4	9
9	8	4	2	1	3	6	7	5
7	5	3	6	9	4	1	8	2
1	2	6	5	8	7	9	3	4

Sudoku # 23

5	6	3	2	4	9	7	8	1
8	1	7	5	3	6	2	9	4
4	9	2	1	8	7	6	3	5
7	3	9	8	6	4	1	5	2
6	2	8	9	5	1	4	7	3
1	4	5	3	7	2	9	6	8
2	8	4	6	9	3	5	1	7
3	7	6	4	1	5	8	2	9
9	5	1	7	2	8	3	4	6

Sudoku # 24

9	1	7	3	8	2	6	5	4
2	4	5	6	1	9	3	8	7
8	6	3	5	7	4	9	1	2
4	7	8	9	2	1	5	3	6
3	2	6	8	5	7	4	9	1
1	5	9	4	3	6	2	7	8
7	3	2	1	4	5	8	6	9
5	9	4	7	6	8	1	2	3
6	8	1	2	9	3	7	4	5

Sudoku # 25

4	2	5	9	8	7	6	1	3
7	6	8	1	4	3	5	2	9
9	1	3	6	2	5	8	7	4
6	3	4	7	9	8	2	5	1
8	5	1	4	6	2	9	3	7
2	7	9	5	3	1	4	8	6
5	8	6	3	1	9	7	4	2
3	4	2	8	7	6	1	9	5
1	9	7	2	5	4	3	6	8

Sudoku # 26

4	3	9	5	6	8	2	7	1
5	6	1	2	9	7	3	4	8
8	7	2	4	3	1	9	5	6
1	8	4	9	7	6	5	2	3
9	2	6	8	5	3	7	1	4
7	5	3	1	2	4	8	6	9
2	4	8	7	1	9	6	3	5
3	1	5	6	8	2	4	9	7
6	9	7	3	4	5	1	8	2

Sudoku # 27

1	5	4	3	8	2	6	9	7
7	6	8	4	5	9	2	3	1
3	9	2	7	6	1	4	5	8
8	3	7	1	4	6	9	2	5
9	4	1	2	7	5	8	6	3
5	2	6	9	3	8	7	1	4
6	1	3	8	2	4	5	7	9
2	8	9	5	1	7	3	4	6
4	7	5	6	9	3	1	8	2

Sudoku # 28

6	2	4	3	5	8	1	7	9
1	7	8	4	9	2	3	6	5
3	5	9	7	6	1	4	8	2
4	9	5	2	3	7	8	1	6
2	3	1	5	8	6	9	4	7
8	6	7	1	4	9	2	5	3
5	8	3	6	2	4	7	9	1
9	1	6	8	7	3	5	2	4
7	4	2	9	1	5	6	3	8

Sudoku # 29

6	2	1	4	9	7	5	8	3
3	9	7	5	8	6	4	1	2
5	4	8	3	1	2	7	9	6
1	6	5	9	4	8	3	2	7
7	8	2	1	6	3	9	4	5
9	3	4	2	7	5	1	6	8
4	7	6	8	5	9	2	3	1
2	5	9	6	3	1	8	7	4
8	1	3	7	2	4	6	5	9

Sudoku # 30

4	6	5	9	8	3	1	7	2
3	9	1	5	7	2	8	4	6
7	2	8	6	4	1	3	5	9
2	3	6	8	9	7	4	1	5
9	8	4	3	1	5	6	2	7
1	5	7	4	2	6	9	3	8
6	1	3	2	5	8	7	9	4
5	7	9	1	6	4	2	8	3
8	4	2	7	3	9	5	6	1

Sudoku # 31

2	3	1	8	7	4	5	9	6
6	7	9	5	2	1	4	8	3
5	4	8	6	9	3	1	2	7
9	1	3	2	5	7	8	6	4
4	5	6	1	3	8	2	7	9
8	2	7	4	6	9	3	5	1
7	9	2	3	4	5	6	1	8
1	6	4	7	8	2	9	3	5
3	8	5	9	1	6	7	4	2

Sudoku # 32

6	2	5	9	3	7	4	8	1
8	3	4	6	1	5	9	7	2
7	1	9	8	4	2	6	5	3
5	9	8	2	6	3	1	4	7
2	7	1	5	9	4	3	6	8
4	6	3	7	8	1	5	2	9
1	4	2	3	7	6	8	9	5
9	5	6	1	2	8	7	3	4
3	8	7	4	5	9	2	1	6

Sudoku # 33

4	9	1	2	6	7	8	3	5
6	5	3	1	8	4	7	9	2
2	8	7	9	5	3	4	6	1
5	3	9	7	1	6	2	8	4
7	4	8	3	9	2	5	1	6
1	2	6	8	4	5	9	7	3
9	1	4	6	2	8	3	5	7
3	6	2	5	7	9	1	4	8
8	7	5	4	3	1	6	2	9

Sudoku # 34

1	8	7	9	4	5	6	3	2
6	4	5	1	3	2	9	7	8
3	2	9	8	6	7	4	5	1
9	7	2	4	8	6	5	1	3
8	1	6	7	5	3	2	4	9
5	3	4	2	9	1	7	8	6
4	9	1	6	7	8	3	2	5
7	5	8	3	2	9	1	6	4
2	6	3	5	1	4	8	9	7

Sudoku # 35

5	3	4	6	7	8	2	1	9
9	1	2	4	5	3	7	8	6
7	8	6	1	9	2	3	5	4
4	7	1	3	2	9	8	6	5
3	2	9	5	8	6	1	4	7
6	5	8	7	1	4	9	2	3
8	4	7	9	6	1	5	3	2
1	9	3	2	4	5	6	7	8
2	6	5	8	3	7	4	9	1

Sudoku # 36

4	6	9	8	2	7	3	1	5
7	1	8	6	3	5	2	4	9
3	5	2	9	1	4	8	7	6
1	3	7	4	5	6	9	8	2
9	4	5	2	7	8	6	3	1
8	2	6	3	9	1	7	5	4
2	9	4	5	8	3	1	6	7
6	8	1	7	4	2	5	9	3
5	7	3	1	6	9	4	2	8

Sudoku # 37

2	9	3	6	1	8	5	7	4
7	1	6	4	5	2	3	9	8
5	4	8	9	7	3	2	6	1
9	8	5	3	6	7	1	4	2
6	7	4	8	2	1	9	5	3
3	2	1	5	4	9	7	8	6
8	6	7	2	3	5	4	1	9
4	5	2	1	9	6	8	3	7
1	3	9	7	8	4	6	2	5

Sudoku # 38

9	3	5	1	2	4	6	8	7
2	6	4	9	7	8	3	5	1
8	1	7	3	6	5	2	4	9
3	7	9	5	1	6	8	2	4
4	8	2	7	3	9	5	1	6
6	5	1	4	8	2	9	7	3
5	4	6	8	9	7	1	3	2
1	9	8	2	4	3	7	6	5
7	2	3	6	5	1	4	9	8

Sudoku # 39

4	5	3	2	7	9	6	8	1
6	2	9	1	4	8	3	5	7
1	7	8	5	3	6	9	4	2
9	4	6	7	2	1	8	3	5
3	1	5	9	8	4	2	7	6
2	8	7	3	6	5	1	9	4
7	6	1	4	9	3	5	2	8
5	9	2	8	1	7	4	6	3
8	3	4	6	5	2	7	1	9

Sudoku # 40

1	3	6	2	5	8	7	9	4
7	9	8	4	3	1	5	6	2
4	5	2	6	7	9	3	8	1
6	7	9	3	8	4	1	2	5
3	2	1	7	9	5	8	4	6
5	8	4	1	6	2	9	3	7
8	1	3	5	2	6	4	7	9
2	4	7	9	1	3	6	5	8
9	6	5	8	4	7	2	1	3

Sudoku # 41

5	8	4	6	2	7	1	9	3
1	9	7	5	3	4	8	2	6
2	3	6	8	1	9	5	4	7
4	1	3	7	6	5	2	8	9
6	2	5	9	8	1	7	3	4
8	7	9	3	4	2	6	1	5
3	4	2	1	5	6	9	7	8
9	5	1	4	7	8	3	6	2
7	6	8	2	9	3	4	5	1

Sudoku # 42

8	2	9	6	1	7	5	4	3
3	7	4	5	2	8	1	9	6
6	5	1	9	4	3	7	8	2
1	6	7	8	3	9	2	5	4
9	3	5	4	7	2	8	6	1
2	4	8	1	6	5	9	3	7
7	9	6	3	5	1	4	2	8
5	1	3	2	8	4	6	7	9
4	8	2	7	9	6	3	1	5

Sudoku # 43

1	3	2	6	7	5	9	4	8
6	4	9	1	3	8	2	7	5
7	5	8	2	4	9	6	1	3
5	6	3	4	8	2	7	9	1
2	9	7	3	1	6	5	8	4
4	8	1	9	5	7	3	6	2
8	1	6	7	2	3	4	5	9
3	7	5	8	9	4	1	2	6
9	2	4	5	6	1	8	3	7

Sudoku # 44

8	3	4	2	9	5	1	7	6
9	5	1	8	6	7	3	4	2
7	2	6	3	4	1	8	5	9
4	7	5	9	2	8	6	3	1
3	8	9	4	1	6	5	2	7
1	6	2	5	7	3	9	8	4
5	4	7	6	8	9	2	1	3
6	1	3	7	5	2	4	9	8
2	9	8	1	3	4	7	6	5

Sudoku # 45

4	6	8	7	3	5	9	1	2
5	9	7	4	1	2	8	6	3
3	1	2	6	8	9	5	4	7
7	5	6	2	9	3	4	8	1
9	4	3	8	7	1	2	5	6
2	8	1	5	4	6	7	3	9
8	7	9	3	6	4	1	2	5
6	2	4	1	5	7	3	9	8
1	3	5	9	2	8	6	7	4

Sudoku # 46

4	9	2	3	5	6	1	8	7
7	5	8	2	1	9	3	4	6
1	3	6	7	4	8	5	2	9
5	1	7	8	3	2	9	6	4
6	2	9	1	7	4	8	5	3
8	4	3	9	6	5	7	1	2
3	6	4	5	8	7	2	9	1
2	7	5	6	9	1	4	3	8
9	8	1	4	2	3	6	7	5

Sudoku # 47

9	3	7	8	6	1	5	2	4
4	8	1	3	2	5	7	6	9
6	2	5	9	4	7	1	3	8
1	7	3	6	8	4	2	9	5
8	4	9	5	1	2	3	7	6
2	5	6	7	9	3	8	4	1
3	1	4	2	5	9	6	8	7
5	6	2	4	7	8	9	1	3
7	9	8	1	3	6	4	5	2

Sudoku # 48

5	9	8	3	4	6	7	1	2
1	3	7	5	2	8	9	6	4
2	6	4	1	9	7	3	8	5
9	2	6	4	8	5	1	7	3
7	5	3	6	1	9	4	2	8
4	8	1	2	7	3	6	5	9
8	4	5	9	6	1	2	3	7
6	7	9	8	3	2	5	4	1
3	1	2	7	5	4	8	9	6

Sudoku # 49

8	4	6	5	1	3	9	7	2
5	1	3	7	9	2	4	6	8
9	7	2	4	8	6	5	3	1
6	2	5	8	3	7	1	4	9
1	3	8	6	4	9	2	5	7
7	9	4	1	2	5	3	8	6
4	6	7	9	5	1	8	2	3
3	5	1	2	6	8	7	9	4
2	8	9	3	7	4	6	1	5

Sudoku # 50

7	1	3	8	4	9	2	6	5
8	5	2	3	6	1	9	4	7
6	4	9	7	2	5	8	3	1
9	8	7	4	5	2	6	1	3
2	3	1	9	7	6	5	8	4
5	6	4	1	8	3	7	2	9
4	7	6	5	1	8	3	9	2
3	2	5	6	9	4	1	7	8
1	9	8	2	3	7	4	5	6

Sudoku # 51

8	4	9	5	3	1	6	7	2
3	2	6	7	9	8	4	5	1
1	7	5	4	2	6	3	8	9
7	6	2	9	1	3	8	4	5
5	8	4	6	7	2	1	9	3
9	1	3	8	4	5	7	2	6
6	3	7	2	5	4	9	1	8
2	9	8	1	6	7	5	3	4
4	5	1	3	8	9	2	6	7

Sudoku # 52

2	3	8	4	1	7	6	9	5
5	6	9	8	3	2	7	1	4
4	7	1	6	5	9	3	2	8
9	4	5	1	2	3	8	6	7
1	2	6	5	7	8	9	4	3
7	8	3	9	4	6	1	5	2
3	9	2	7	6	5	4	8	1
6	1	7	2	8	4	5	3	9
8	5	4	3	9	1	2	7	6

Sudoku # 53

4	9	2	7	6	5	8	3	1
3	8	6	2	1	4	5	7	9
5	1	7	3	8	9	2	6	4
6	7	8	5	3	1	9	4	2
9	4	1	6	2	8	3	5	7
2	5	3	9	4	7	1	8	6
1	3	5	4	7	2	6	9	8
7	2	9	8	5	6	4	1	3
8	6	4	1	9	3	7	2	5

Sudoku # 54

8	6	2	4	9	3	1	7	5
4	5	7	1	8	2	3	9	6
3	1	9	6	7	5	2	8	4
9	3	8	5	4	7	6	2	1
2	7	5	3	1	6	9	4	8
6	4	1	9	2	8	7	5	3
5	2	6	8	3	9	4	1	7
7	8	4	2	6	1	5	3	9
1	9	3	7	5	4	8	6	2

Sudoku # 55

4	9	2	3	1	7	6	5	8
7	6	1	4	5	8	9	2	3
3	8	5	2	6	9	1	4	7
5	3	6	8	7	2	4	1	9
1	4	8	9	3	6	2	7	5
9	2	7	1	4	5	8	3	6
8	5	4	6	2	3	7	9	1
2	7	9	5	8	1	3	6	4
6	1	3	7	9	4	5	8	2

Sudoku # 56

4	8	7	5	6	9	3	1	2
9	6	5	2	3	1	4	7	8
2	3	1	4	7	8	9	5	6
6	1	8	7	4	3	5	2	9
3	4	2	1	9	5	8	6	7
5	7	9	8	2	6	1	4	3
7	2	3	9	5	4	6	8	1
8	5	6	3	1	2	7	9	4
1	9	4	6	8	7	2	3	5

Sudoku # 57

7	1	3	8	9	5	4	6	2
8	4	2	3	1	6	7	9	5
5	6	9	4	7	2	3	8	1
2	9	8	1	5	7	6	3	4
3	5	6	9	8	4	2	1	7
4	7	1	2	6	3	8	5	9
9	2	5	6	4	8	1	7	3
1	8	4	7	3	9	5	2	6
6	3	7	5	2	1	9	4	8

Sudoku # 58

6	1	9	2	3	5	8	4	7
2	3	5	8	7	4	9	1	6
8	7	4	1	6	9	2	5	3
9	2	3	7	1	6	5	8	4
4	5	1	3	9	8	7	6	2
7	6	8	4	5	2	3	9	1
1	8	6	5	2	7	4	3	9
5	9	2	6	4	3	1	7	8
3	4	7	9	8	1	6	2	5

Sudoku # 59

7	8	6	9	2	4	5	1	3
2	4	9	1	5	3	8	7	6
5	3	1	7	8	6	2	9	4
4	1	2	3	6	8	7	5	9
9	6	3	2	7	5	4	8	1
8	7	5	4	1	9	3	6	2
1	9	8	5	3	2	6	4	7
3	5	4	6	9	7	1	2	8
6	2	7	8	4	1	9	3	5

Sudoku # 60

7	9	4	2	8	1	6	3	5
2	8	3	7	6	5	4	1	9
5	6	1	4	9	3	2	7	8
9	4	7	8	5	6	1	2	3
8	3	6	1	7	2	9	5	4
1	2	5	9	3	4	8	6	7
3	5	2	6	4	9	7	8	1
6	7	9	3	1	8	5	4	2
4	1	8	5	2	7	3	9	6

Sudoku # 61

4	5	1	7	8	9	6	3	2
7	9	8	2	6	3	5	1	4
6	2	3	1	4	5	8	9	7
9	8	6	4	3	2	7	5	1
2	1	5	6	7	8	3	4	9
3	4	7	9	5	1	2	8	6
5	7	4	8	1	6	9	2	3
8	6	9	3	2	4	1	7	5
1	3	2	5	9	7	4	6	8

Sudoku # 62

4	3	5	7	9	1	2	8	6
2	7	9	3	8	6	4	1	5
6	8	1	4	5	2	7	3	9
7	2	6	1	3	5	9	4	8
9	1	8	6	2	4	3	5	7
3	5	4	9	7	8	6	2	1
1	4	3	5	6	7	8	9	2
8	9	7	2	1	3	5	6	4
5	6	2	8	4	9	1	7	3

Sudoku # 63

8	4	3	9	6	2	7	1	5
2	5	9	7	1	3	6	8	4
6	7	1	8	5	4	2	9	3
9	3	4	1	8	6	5	2	7
5	2	6	4	9	7	1	3	8
7	1	8	3	2	5	4	6	9
3	9	5	2	7	1	8	4	6
4	6	2	5	3	8	9	7	1
1	8	7	6	4	9	3	5	2

Sudoku # 64

2	3	4	5	6	7	9	8	1
5	8	6	1	9	2	7	3	4
7	1	9	4	8	3	2	5	6
3	2	1	9	5	6	4	7	8
4	9	5	2	7	8	1	6	3
6	7	8	3	4	1	5	2	9
9	6	2	8	1	5	3	4	7
1	5	7	6	3	4	8	9	2
8	4	3	7	2	9	6	1	5

Sudoku # 65

4	3	1	9	5	8	6	2	7
7	8	9	6	2	3	5	1	4
5	2	6	1	7	4	3	8	9
2	1	8	3	9	5	4	7	6
6	5	4	8	1	7	9	3	2
9	7	3	4	6	2	8	5	1
8	9	7	5	4	1	2	6	3
1	4	5	2	3	6	7	9	8
3	6	2	7	8	9	1	4	5

Sudoku # 66

4	6	3	9	7	2	8	5	1
8	9	7	3	5	1	4	2	6
2	1	5	6	4	8	7	3	9
1	7	8	4	9	5	2	6	3
9	4	6	1	2	3	5	7	8
3	5	2	7	8	6	9	1	4
6	2	4	8	3	7	1	9	5
7	3	9	5	1	4	6	8	2
5	8	1	2	6	9	3	4	7

Sudoku # 67

1	5	9	4	3	6	7	8	2
8	3	6	7	9	2	1	4	5
2	7	4	8	5	1	6	3	9
5	1	8	2	4	7	3	9	6
3	4	2	6	1	9	5	7	8
9	6	7	3	8	5	4	2	1
6	9	3	5	2	4	8	1	7
7	8	1	9	6	3	2	5	4
4	2	5	1	7	8	9	6	3

Sudoku # 68

3	6	1	2	5	7	9	4	8
7	4	2	6	8	9	3	5	1
5	9	8	4	1	3	2	7	6
6	7	4	5	3	1	8	2	9
8	2	9	7	4	6	1	3	5
1	3	5	9	2	8	7	6	4
2	1	3	8	6	4	5	9	7
4	5	7	1	9	2	6	8	3
9	8	6	3	7	5	4	1	2

Sudoku # 69

6	1	7	3	4	9	2	5	8
4	8	3	5	1	2	6	7	9
9	5	2	8	7	6	1	4	3
2	4	8	7	9	3	5	1	6
5	7	9	1	6	4	3	8	2
3	6	1	2	8	5	4	9	7
8	3	5	9	2	1	7	6	4
7	2	6	4	5	8	9	3	1
1	9	4	6	3	7	8	2	5

Sudoku # 70

5	7	3	9	1	8	2	4	6
1	4	2	7	5	6	3	8	9
6	8	9	3	4	2	1	5	7
2	5	7	1	9	4	8	6	3
9	3	4	8	6	5	7	2	1
8	1	6	2	7	3	5	9	4
7	6	5	4	8	1	9	3	2
3	9	8	6	2	7	4	1	5
4	2	1	5	3	9	6	7	8

Sudoku # 71

5	9	4	1	7	8	6	2	3
8	2	1	4	6	3	7	5	9
7	3	6	2	5	9	8	4	1
4	6	8	7	1	5	3	9	2
3	7	9	8	2	4	1	6	5
2	1	5	9	3	6	4	7	8
6	5	2	3	4	1	9	8	7
9	4	3	5	8	7	2	1	6
1	8	7	6	9	2	5	3	4

Sudoku # 72

1	4	6	9	5	2	7	3	8
2	3	9	8	1	7	4	6	5
7	5	8	3	6	4	9	1	2
4	9	1	6	2	3	5	8	7
6	8	3	5	7	9	2	4	1
5	2	7	1	4	8	3	9	6
3	6	2	4	8	5	1	7	9
9	1	5	7	3	6	8	2	4
8	7	4	2	9	1	6	5	3

Sudoku # 73

3	1	8	6	5	4	2	7	9
7	2	5	3	8	9	6	4	1
4	9	6	1	7	2	8	5	3
5	8	1	7	9	3	4	6	2
2	6	7	5	4	1	3	9	8
9	4	3	8	2	6	7	1	5
6	3	4	2	1	5	9	8	7
8	5	2	9	6	7	1	3	4
1	7	9	4	3	8	5	2	6

Sudoku # 74

3	2	9	4	5	8	1	7	6
7	5	6	9	3	1	2	8	4
4	8	1	7	2	6	9	3	5
5	1	4	8	7	3	6	9	2
9	7	2	5	6	4	8	1	3
8	6	3	1	9	2	4	5	7
6	3	8	2	1	7	5	4	9
2	4	5	3	8	9	7	6	1
1	9	7	6	4	5	3	2	8

Sudoku # 75

6	1	2	7	5	3	9	8	4
8	7	5	4	9	6	3	1	2
4	3	9	2	1	8	7	5	6
2	4	1	3	7	9	8	6	5
3	9	8	5	6	1	4	2	7
5	6	7	8	2	4	1	9	3
1	5	4	9	3	2	6	7	8
7	8	6	1	4	5	2	3	9
9	2	3	6	8	7	5	4	1

Sudoku # 76

5	4	3	8	6	1	9	7	2
8	6	2	4	9	7	3	5	1
1	9	7	5	3	2	4	8	6
2	1	4	7	8	6	5	3	9
7	3	5	1	4	9	2	6	8
6	8	9	2	5	3	1	4	7
4	2	1	3	7	8	6	9	5
3	7	6	9	2	5	8	1	4
9	5	8	6	1	4	7	2	3

Sudoku # 77

3	6	4	8	5	9	1	7	2
1	5	8	7	4	2	9	6	3
7	9	2	6	1	3	5	8	4
4	8	3	9	6	7	2	1	5
2	1	6	3	8	5	7	4	9
5	7	9	1	2	4	8	3	6
8	3	5	4	9	1	6	2	7
9	4	1	2	7	6	3	5	8
6	2	7	5	3	8	4	9	1

Sudoku # 78

3	5	1	6	4	8	7	9	2
9	4	8	2	5	7	3	1	6
6	7	2	1	3	9	8	5	4
1	8	5	4	9	3	2	6	7
2	3	6	7	8	1	9	4	5
4	9	7	5	2	6	1	3	8
7	1	3	8	6	4	5	2	9
8	2	4	9	1	5	6	7	3
5	6	9	3	7	2	4	8	1

Sudoku # 79

6	8	2	3	1	5	9	7	4
9	4	7	8	2	6	1	3	5
3	1	5	4	7	9	8	2	6
1	5	4	6	8	3	7	9	2
2	6	8	1	9	7	5	4	3
7	9	3	5	4	2	6	8	1
5	3	9	2	6	8	4	1	7
4	7	6	9	3	1	2	5	8
8	2	1	7	5	4	3	6	9

Sudoku # 80

9	3	4	5	7	2	1	6	8
1	7	2	3	8	6	9	5	4
8	5	6	1	9	4	2	3	7
2	4	9	7	6	3	8	1	5
5	1	7	2	4	8	6	9	3
3	6	8	9	1	5	4	7	2
7	9	3	4	2	1	5	8	6
6	2	1	8	5	7	3	4	9
4	8	5	6	3	9	7	2	1

Sudoku # 81

9	3	5	1	4	2	7	8	6
8	1	7	9	6	5	2	3	4
2	6	4	8	7	3	9	5	1
3	2	8	4	1	6	5	7	9
7	5	1	3	2	9	4	6	8
4	9	6	5	8	7	1	2	3
5	8	9	7	3	1	6	4	2
1	4	2	6	5	8	3	9	7
6	7	3	2	9	4	8	1	5

Sudoku # 82

9	5	6	7	3	2	1	4	8
8	2	4	5	1	6	9	7	3
1	7	3	8	4	9	5	2	6
7	1	5	9	6	4	8	3	2
2	3	9	1	7	8	6	5	4
6	4	8	2	5	3	7	9	1
5	9	2	4	8	1	3	6	7
4	6	1	3	9	7	2	8	5
3	8	7	6	2	5	4	1	9

Sudoku # 83

5	8	3	4	7	2	6	9	1
7	6	1	3	8	9	5	2	4
2	9	4	5	6	1	3	8	7
8	7	2	9	3	6	4	1	5
9	3	6	1	4	5	2	7	8
1	4	5	7	2	8	9	3	6
3	1	9	8	5	4	7	6	2
4	2	7	6	1	3	8	5	9
6	5	8	2	9	7	1	4	3

Sudoku # 84

9	6	1	3	2	8	5	4	7
5	4	3	1	6	7	9	2	8
7	2	8	4	9	5	6	1	3
2	3	9	5	4	6	7	8	1
6	5	4	7	8	1	3	9	2
8	1	7	2	3	9	4	6	5
3	9	6	8	7	2	1	5	4
4	8	5	9	1	3	2	7	6
1	7	2	6	5	4	8	3	9

Sudoku # 85

1	2	6	4	9	5	3	7	8
3	4	5	8	6	7	2	9	1
9	7	8	2	3	1	5	4	6
4	9	7	1	2	3	6	8	5
6	8	3	9	5	4	1	2	7
5	1	2	6	7	8	9	3	4
8	3	9	7	1	6	4	5	2
7	5	1	3	4	2	8	6	9
2	6	4	5	8	9	7	1	3

Sudoku # 86

2	6	5	1	3	7	4	9	8
4	7	1	5	9	8	6	3	2
9	8	3	4	6	2	1	7	5
5	9	6	8	2	1	3	4	7
3	4	2	6	7	9	5	8	1
7	1	8	3	4	5	2	6	9
1	3	7	9	5	4	8	2	6
8	2	4	7	1	6	9	5	3
6	5	9	2	8	3	7	1	4

Sudoku # 87

6	3	7	8	9	5	2	1	4
1	5	9	4	2	7	3	6	8
4	8	2	3	6	1	5	7	9
2	9	3	6	4	8	1	5	7
5	6	4	7	1	3	8	9	2
8	7	1	9	5	2	4	3	6
9	2	6	5	3	4	7	8	1
3	4	8	1	7	9	6	2	5
7	1	5	2	8	6	9	4	3

Sudoku # 88

6	4	5	8	1	7	9	2	3
8	7	1	2	3	9	4	6	5
9	2	3	5	6	4	8	1	7
4	1	8	3	5	6	7	9	2
5	3	6	7	9	2	1	8	4
7	9	2	4	8	1	5	3	6
1	6	7	9	4	3	2	5	8
3	5	4	1	2	8	6	7	9
2	8	9	6	7	5	3	4	1

Sudoku # 89

1	6	5	3	8	7	9	2	4
4	7	3	6	2	9	8	1	5
2	9	8	1	4	5	3	6	7
6	1	7	4	3	2	5	8	9
9	5	2	8	7	1	6	4	3
3	8	4	5	9	6	2	7	1
5	4	1	9	6	8	7	3	2
8	2	9	7	1	3	4	5	6
7	3	6	2	5	4	1	9	8

Sudoku # 90

2	6	8	3	1	4	9	5	7
4	3	1	5	9	7	2	6	8
7	5	9	2	6	8	3	1	4
5	8	4	6	7	2	1	3	9
9	7	2	4	3	1	6	8	5
6	1	3	9	8	5	7	4	2
3	2	5	7	4	6	8	9	1
8	4	6	1	2	9	5	7	3
1	9	7	8	5	3	4	2	6

Sudoku # 91

2	4	8	6	3	7	5	1	9
9	1	7	5	2	4	3	8	6
3	5	6	1	8	9	7	4	2
4	6	1	7	5	2	9	3	8
5	2	9	3	1	8	4	6	7
8	7	3	4	9	6	1	2	5
6	3	2	9	4	5	8	7	1
7	9	4	8	6	1	2	5	3
1	8	5	2	7	3	6	9	4

Sudoku # 92

5	9	6	3	1	4	2	7	8
3	7	8	5	9	2	4	6	1
1	4	2	8	7	6	5	3	9
8	3	5	4	2	1	6	9	7
7	1	9	6	5	8	3	4	2
6	2	4	9	3	7	1	8	5
4	5	3	2	8	9	7	1	6
9	6	1	7	4	5	8	2	3
2	8	7	1	6	3	9	5	4

Sudoku # 93

3	2	8	1	6	5	7	9	4
7	5	4	9	8	2	1	6	3
1	9	6	7	4	3	8	2	5
5	7	2	3	9	4	6	8	1
6	1	9	2	5	8	3	4	7
8	4	3	6	7	1	2	5	9
4	3	1	8	2	9	5	7	6
9	8	7	5	3	6	4	1	2
2	6	5	4	1	7	9	3	8

Sudoku # 94

9	5	2	8	7	1	3	6	4
8	3	4	5	9	6	1	7	2
6	7	1	3	4	2	5	9	8
1	2	6	7	8	9	4	3	5
3	9	7	4	1	5	2	8	6
5	4	8	6	2	3	7	1	9
4	8	9	2	3	7	6	5	1
2	6	3	1	5	8	9	4	7
7	1	5	9	6	4	8	2	3

Sudoku # 95

6	2	3	1	8	9	4	5	7
7	8	4	2	3	5	1	6	9
9	1	5	6	4	7	3	2	8
2	5	8	9	7	4	6	1	3
1	6	9	3	2	8	7	4	5
3	4	7	5	1	6	8	9	2
8	7	2	4	5	1	9	3	6
5	9	1	8	6	3	2	7	4
4	3	6	7	9	2	5	8	1

Sudoku # 96

7	6	3	4	8	2	1	9	5
1	8	9	7	5	6	4	2	3
5	2	4	9	3	1	8	7	6
3	4	6	5	2	9	7	8	1
9	5	8	6	1	7	3	4	2
2	1	7	8	4	3	5	6	9
4	7	2	3	6	5	9	1	8
8	3	1	2	9	4	6	5	7
6	9	5	1	7	8	2	3	4

Sudoku # 97

4	5	9	8	1	7	3	6	2
6	1	7	3	2	9	4	8	5
8	2	3	4	5	6	7	9	1
3	8	4	6	9	5	2	1	7
5	7	1	2	3	8	9	4	6
9	6	2	1	7	4	8	5	3
1	9	6	7	8	2	5	3	4
2	4	5	9	6	3	1	7	8
7	3	8	5	4	1	6	2	9

Sudoku # 98

1	8	4	7	2	6	3	5	9
2	3	5	9	1	8	6	7	4
6	7	9	3	4	5	1	8	2
3	9	8	1	5	4	2	6	7
5	6	2	8	9	7	4	3	1
4	1	7	2	6	3	8	9	5
8	5	6	4	7	2	9	1	3
7	4	1	6	3	9	5	2	8
9	2	3	5	8	1	7	4	6

Sudoku # 99

2	7	8	3	6	5	4	9	1
1	6	5	2	4	9	3	8	7
9	3	4	8	7	1	2	5	6
6	4	7	5	1	8	9	3	2
5	8	2	7	9	3	1	6	4
3	9	1	4	2	6	8	7	5
8	5	6	1	3	2	7	4	9
7	2	3	9	5	4	6	1	8
4	1	9	6	8	7	5	2	3

Sudoku # 100

2	9	5	8	6	4	1	7	3
8	7	4	3	2	1	9	6	5
6	3	1	5	9	7	8	2	4
5	8	6	1	3	2	4	9	7
7	2	3	4	8	9	6	5	1
1	4	9	6	7	5	2	3	8
4	5	2	7	1	6	3	8	9
3	6	7	9	4	8	5	1	2
9	1	8	2	5	3	7	4	6

Sudoku # 101

7	1	5	8	6	9	3	4	2
6	9	2	4	1	3	8	5	7
3	8	4	5	7	2	1	6	9
9	2	1	6	8	7	5	3	4
5	4	6	9	3	1	2	7	8
8	7	3	2	5	4	6	9	1
2	3	8	7	9	6	4	1	5
4	6	9	1	2	5	7	8	3
1	5	7	3	4	8	9	2	6

Sudoku # 102

9	7	5	3	4	1	8	6	2
2	1	3	8	7	6	9	4	5
8	6	4	2	5	9	7	1	3
1	5	2	6	9	8	4	3	7
6	4	8	7	2	3	5	9	1
3	9	7	5	1	4	6	2	8
4	2	6	1	8	5	3	7	9
5	3	1	9	6	7	2	8	4
7	8	9	4	3	2	1	5	6

Sudoku # 103

3	1	2	5	6	7	8	4	9
5	4	8	9	1	2	6	7	3
9	6	7	8	3	4	5	1	2
4	8	6	3	2	9	1	5	7
7	3	5	4	8	1	9	2	6
1	2	9	7	5	6	4	3	8
6	7	1	2	4	8	3	9	5
8	9	3	1	7	5	2	6	4
2	5	4	6	9	3	7	8	1

Sudoku # 104

1	8	9	5	7	6	2	3	4
6	2	3	9	1	4	8	7	5
5	7	4	8	3	2	6	1	9
2	4	6	7	5	1	9	8	3
7	9	8	2	4	3	1	5	6
3	1	5	6	9	8	7	4	2
9	5	1	4	2	7	3	6	8
8	3	2	1	6	5	4	9	7
4	6	7	3	8	9	5	2	1

Sudoku # 105

8	4	3	9	2	7	5	1	6
6	5	2	8	1	4	3	9	7
7	9	1	6	5	3	8	2	4
2	6	4	3	8	1	7	5	9
9	3	5	7	6	2	1	4	8
1	8	7	4	9	5	6	3	2
4	1	9	5	7	6	2	8	3
5	7	8	2	3	9	4	6	1
3	2	6	1	4	8	9	7	5

Sudoku # 106

5	1	8	4	9	7	2	3	6
7	2	4	5	3	6	8	1	9
6	9	3	1	2	8	7	4	5
1	8	7	9	5	3	4	6	2
3	6	9	7	4	2	5	8	1
4	5	2	6	8	1	3	9	7
9	4	1	3	7	5	6	2	8
2	3	5	8	6	9	1	7	4
8	7	6	2	1	4	9	5	3

Sudoku # 107

2	3	1	6	5	4	9	7	8
7	8	5	2	9	1	4	3	6
4	6	9	3	8	7	5	1	2
8	1	4	5	3	9	6	2	7
9	7	6	4	2	8	1	5	3
3	5	2	7	1	6	8	4	9
6	4	8	1	7	3	2	9	5
1	2	3	9	6	5	7	8	4
5	9	7	8	4	2	3	6	1

Sudoku # 108

2	1	3	7	8	4	6	9	5
8	6	5	1	9	2	3	7	4
4	7	9	6	3	5	8	2	1
9	2	6	3	4	7	1	5	8
5	4	8	9	2	1	7	6	3
7	3	1	8	5	6	9	4	2
3	5	7	4	1	9	2	8	6
1	9	2	5	6	8	4	3	7
6	8	4	2	7	3	5	1	9

Sudoku # 109

6	5	9	2	8	4	1	7	3
8	2	7	6	3	1	5	9	4
3	4	1	7	5	9	2	8	6
4	8	2	5	9	6	3	1	7
7	1	5	8	4	3	9	6	2
9	6	3	1	2	7	8	4	5
2	7	8	4	1	5	6	3	9
1	3	4	9	6	2	7	5	8
5	9	6	3	7	8	4	2	1

Sudoku # 110

8	6	7	5	2	3	4	9	1
5	4	2	8	1	9	6	7	3
1	9	3	7	6	4	5	2	8
9	8	6	3	5	2	7	1	4
4	7	1	6	9	8	2	3	5
2	3	5	4	7	1	9	8	6
6	2	4	1	3	7	8	5	9
3	5	9	2	8	6	1	4	7
7	1	8	9	4	5	3	6	2

Sudoku # 111

2	6	9	4	1	5	3	7	8
1	4	5	8	3	7	9	6	2
8	3	7	9	2	6	4	1	5
4	8	3	5	7	2	6	9	1
5	1	6	3	9	8	7	2	4
7	9	2	6	4	1	5	8	3
6	5	4	2	8	9	1	3	7
9	7	8	1	5	3	2	4	6
3	2	1	7	6	4	8	5	9

Sudoku # 112

3	6	5	7	2	1	8	9	4
2	9	8	5	6	4	1	7	3
1	4	7	9	3	8	5	6	2
9	1	2	6	7	3	4	8	5
6	8	4	2	1	5	7	3	9
5	7	3	8	4	9	2	1	6
8	2	9	1	5	6	3	4	7
7	3	6	4	8	2	9	5	1
4	5	1	3	9	7	6	2	8

Sudoku # 113

8	6	1	3	4	5	9	7	2
2	3	9	7	6	1	5	4	8
5	4	7	2	8	9	3	6	1
4	2	3	6	5	7	1	8	9
1	8	5	9	3	4	6	2	7
7	9	6	1	2	8	4	3	5
9	1	4	8	7	6	2	5	3
3	5	8	4	1	2	7	9	6
6	7	2	5	9	3	8	1	4

Sudoku # 114

8	2	4	6	7	1	3	9	5
7	1	6	3	5	9	8	4	2
3	9	5	8	2	4	7	6	1
6	7	8	9	4	2	1	5	3
2	4	3	1	6	5	9	7	8
1	5	9	7	8	3	4	2	6
4	6	1	5	3	7	2	8	9
9	8	7	2	1	6	5	3	4
5	3	2	4	9	8	6	1	7

Sudoku # 115

6	8	2	7	5	1	3	4	9
7	5	3	2	4	9	8	1	6
4	1	9	6	8	3	5	2	7
3	7	5	9	2	6	4	8	1
1	4	6	3	7	8	2	9	5
9	2	8	4	1	5	7	6	3
8	3	1	5	9	4	6	7	2
2	6	4	1	3	7	9	5	8
5	9	7	8	6	2	1	3	4

Sudoku # 116

3	2	1	9	7	8	4	6	5
9	5	8	2	4	6	1	7	3
6	7	4	5	1	3	2	9	8
7	1	6	4	3	9	5	8	2
5	3	2	7	8	1	9	4	6
4	8	9	6	5	2	7	3	1
1	6	7	3	9	5	8	2	4
2	9	5	8	6	4	3	1	7
8	4	3	1	2	7	6	5	9

Sudoku # 117

9	7	2	1	4	5	8	6	3
4	6	5	7	8	3	9	2	1
1	8	3	6	2	9	7	4	5
3	1	4	2	9	8	6	5	7
7	2	8	4	5	6	3	1	9
6	5	9	3	7	1	2	8	4
5	9	6	8	1	7	4	3	2
2	3	7	5	6	4	1	9	8
8	4	1	9	3	2	5	7	6

Sudoku # 118

4	6	2	9	1	3	5	7	8
8	9	7	5	6	4	3	1	2
1	3	5	7	8	2	4	6	9
9	2	3	1	4	7	8	5	6
7	1	6	3	5	8	2	9	4
5	4	8	6	2	9	7	3	1
3	8	1	4	7	6	9	2	5
6	7	4	2	9	5	1	8	3
2	5	9	8	3	1	6	4	7

Sudoku # 119

8	6	7	3	9	2	5	1	4
2	3	9	5	4	1	7	8	6
4	5	1	6	7	8	3	2	9
1	7	2	8	6	3	9	4	5
3	4	5	9	1	7	2	6	8
9	8	6	4	2	5	1	7	3
6	2	3	7	5	4	8	9	1
5	1	4	2	8	9	6	3	7
7	9	8	1	3	6	4	5	2

Sudoku # 120

6	1	4	8	9	7	5	2	3
9	3	2	1	5	4	7	8	6
7	8	5	3	2	6	9	1	4
2	4	7	6	3	8	1	9	5
5	6	1	2	7	9	4	3	8
8	9	3	4	1	5	2	6	7
3	2	6	5	4	1	8	7	9
1	5	9	7	8	3	6	4	2
4	7	8	9	6	2	3	5	1

Sudoku # 121

3	6	7	4	9	1	8	2	5
9	8	4	6	2	5	7	1	3
1	2	5	3	8	7	9	4	6
5	7	9	8	3	4	1	6	2
4	1	2	5	6	9	3	8	7
8	3	6	7	1	2	5	9	4
6	9	3	2	5	8	4	7	1
7	5	8	1	4	6	2	3	9
2	4	1	9	7	3	6	5	8

Sudoku # 122

8	3	2	4	6	9	5	1	7
6	7	1	3	2	5	9	4	8
5	9	4	7	1	8	6	3	2
9	1	7	5	4	6	2	8	3
4	5	8	9	3	2	1	7	6
2	6	3	1	8	7	4	9	5
3	4	6	2	7	1	8	5	9
1	8	9	6	5	3	7	2	4
7	2	5	8	9	4	3	6	1

Sudoku # 123

9	4	2	5	7	3	1	6	8
7	6	1	8	4	9	2	3	5
5	3	8	6	2	1	9	7	4
3	9	7	4	6	2	8	5	1
8	2	5	1	3	7	6	4	9
4	1	6	9	8	5	3	2	7
6	8	9	3	5	4	7	1	2
2	5	3	7	1	8	4	9	6
1	7	4	2	9	6	5	8	3

Sudoku # 124

4	2	7	9	6	8	5	3	1
9	8	1	7	3	5	6	2	4
6	3	5	2	4	1	7	9	8
1	4	2	6	7	9	3	8	5
7	6	3	8	5	4	2	1	9
8	5	9	1	2	3	4	6	7
5	1	6	3	9	7	8	4	2
2	7	8	4	1	6	9	5	3
3	9	4	5	8	2	1	7	6

Sudoku # 125

1	5	6	3	4	8	9	2	7
4	9	3	2	5	7	1	8	6
7	8	2	6	1	9	3	5	4
8	3	1	5	2	4	6	7	9
9	7	4	1	8	6	2	3	5
6	2	5	7	9	3	4	1	8
3	1	9	4	7	5	8	6	2
2	4	7	8	6	1	5	9	3
5	6	8	9	3	2	7	4	1

Sudoku # 126

9	6	3	7	5	4	1	8	2
4	1	2	3	6	8	9	7	5
5	7	8	1	9	2	3	4	6
8	9	5	2	3	1	4	6	7
2	4	6	5	7	9	8	1	3
7	3	1	8	4	6	2	5	9
3	8	4	6	2	7	5	9	1
6	2	9	4	1	5	7	3	8
1	5	7	9	8	3	6	2	4

Sudoku # 127

7	5	6	2	3	4	8	9	1
1	9	4	8	7	6	5	2	3
2	8	3	9	1	5	7	4	6
5	3	2	1	6	9	4	7	8
9	6	8	3	4	7	1	5	2
4	7	1	5	8	2	6	3	9
8	1	7	4	9	3	2	6	5
3	4	5	6	2	1	9	8	7
6	2	9	7	5	8	3	1	4

Sudoku # 128

5	9	4	7	8	6	3	2	1
2	6	1	3	9	4	5	8	7
3	8	7	5	2	1	6	4	9
9	3	5	2	7	8	4	1	6
4	2	6	9	1	5	7	3	8
1	7	8	6	4	3	2	9	5
6	1	9	4	3	7	8	5	2
8	5	3	1	6	2	9	7	4
7	4	2	8	5	9	1	6	3

Sudoku # 129

3	6	8	5	4	1	9	7	2
1	9	4	7	8	2	3	5	6
2	7	5	6	3	9	8	1	4
6	5	3	4	9	8	1	2	7
4	1	9	3	2	7	6	8	5
7	8	2	1	5	6	4	9	3
8	2	7	9	6	4	5	3	1
9	3	6	2	1	5	7	4	8
5	4	1	8	7	3	2	6	9

Sudoku # 130

1	2	5	7	9	3	4	6	8
7	9	6	5	8	4	1	3	2
4	8	3	1	2	6	9	7	5
3	4	1	2	5	8	6	9	7
5	7	2	9	6	1	3	8	4
8	6	9	4	3	7	5	2	1
2	5	8	6	4	9	7	1	3
9	1	4	3	7	2	8	5	6
6	3	7	8	1	5	2	4	9

Sudoku # 131

5	4	6	2	1	8	9	7	3
1	3	8	5	9	7	4	6	2
7	2	9	6	4	3	8	1	5
4	7	1	3	2	6	5	8	9
8	9	2	4	5	1	7	3	6
6	5	3	7	8	9	2	4	1
9	1	7	8	3	5	6	2	4
3	8	4	9	6	2	1	5	7
2	6	5	1	7	4	3	9	8

Sudoku # 132

5	9	3	6	1	8	2	7	4
6	2	4	7	3	5	1	8	9
8	1	7	9	2	4	5	3	6
3	7	9	5	8	6	4	1	2
4	6	2	3	9	1	8	5	7
1	5	8	2	4	7	9	6	3
2	8	1	4	6	3	7	9	5
7	4	6	1	5	9	3	2	8
9	3	5	8	7	2	6	4	1

Sudoku # 133

4	7	6	5	8	9	3	2	1
2	9	3	4	7	1	5	6	8
1	8	5	2	6	3	4	9	7
9	2	7	8	5	6	1	3	4
3	6	4	7	1	2	8	5	9
5	1	8	9	3	4	2	7	6
8	5	9	3	4	7	6	1	2
7	3	1	6	2	8	9	4	5
6	4	2	1	9	5	7	8	3

Sudoku # 134

9	7	1	4	2	5	3	8	6
6	3	5	7	8	1	4	9	2
8	2	4	3	9	6	1	5	7
7	6	2	9	4	3	5	1	8
4	1	3	6	5	8	2	7	9
5	8	9	1	7	2	6	4	3
3	5	6	8	1	7	9	2	4
2	4	7	5	3	9	8	6	1
1	9	8	2	6	4	7	3	5

Sudoku # 135

2	8	3	6	4	5	9	7	1
4	7	9	3	8	1	5	2	6
1	6	5	9	2	7	8	3	4
3	1	4	2	5	8	6	9	7
7	2	8	4	9	6	3	1	5
9	5	6	1	7	3	4	8	2
5	4	1	7	3	9	2	6	8
6	3	2	8	1	4	7	5	9
8	9	7	5	6	2	1	4	3

Sudoku # 136

9	7	4	5	6	3	8	1	2
1	2	3	8	7	4	5	6	9
5	8	6	9	1	2	4	3	7
4	6	7	3	8	9	2	5	1
3	5	1	6	2	7	9	8	4
2	9	8	1	4	5	3	7	6
8	3	2	7	9	6	1	4	5
7	4	5	2	3	1	6	9	8
6	1	9	4	5	8	7	2	3

Sudoku # 137

8	3	2	4	9	5	7	1	6
7	6	4	2	8	1	9	5	3
1	5	9	3	6	7	2	8	4
4	7	1	9	3	8	6	2	5
2	8	6	5	1	4	3	7	9
3	9	5	6	7	2	1	4	8
6	4	8	7	2	9	5	3	1
9	1	7	8	5	3	4	6	2
5	2	3	1	4	6	8	9	7

Sudoku # 138

5	8	4	1	3	7	2	6	9
2	6	9	8	4	5	3	7	1
3	7	1	6	2	9	5	4	8
9	2	3	4	7	1	6	8	5
8	1	7	2	5	6	4	9	3
6	4	5	3	9	8	1	2	7
1	9	6	5	8	4	7	3	2
4	3	8	7	1	2	9	5	6
7	5	2	9	6	3	8	1	4

Sudoku # 139

8	2	6	7	3	4	9	1	5
7	9	3	8	1	5	4	6	2
5	1	4	6	9	2	7	3	8
6	5	2	9	4	1	3	8	7
3	4	7	5	2	8	1	9	6
9	8	1	3	6	7	5	2	4
2	7	8	1	5	9	6	4	3
1	3	5	4	8	6	2	7	9
4	6	9	2	7	3	8	5	1

Sudoku # 140

8	5	6	1	4	7	9	3	2
7	1	9	6	3	2	4	8	5
3	4	2	8	5	9	1	6	7
4	7	8	5	1	3	6	2	9
5	9	1	4	2	6	8	7	3
6	2	3	7	9	8	5	1	4
9	6	7	3	8	5	2	4	1
2	3	4	9	6	1	7	5	8
1	8	5	2	7	4	3	9	6

Sudoku # 141

4	2	5	9	6	8	1	3	7
3	1	8	4	7	5	2	9	6
9	7	6	1	3	2	5	4	8
6	5	2	3	4	9	8	7	1
1	4	7	2	8	6	9	5	3
8	9	3	7	5	1	4	6	2
2	6	1	5	9	3	7	8	4
7	8	9	6	1	4	3	2	5
5	3	4	8	2	7	6	1	9

Sudoku # 142

1	4	8	2	7	5	3	6	9
7	3	2	1	9	6	8	4	5
5	6	9	3	8	4	2	1	7
4	8	6	7	3	2	9	5	1
9	2	1	6	5	8	7	3	4
3	7	5	9	4	1	6	8	2
8	5	7	4	6	9	1	2	3
6	1	3	5	2	7	4	9	8
2	9	4	8	1	3	5	7	6

Sudoku # 143

4	8	6	5	3	2	9	1	7
3	5	1	8	9	7	4	2	6
7	9	2	4	1	6	5	3	8
2	6	8	3	5	9	1	7	4
5	7	4	6	2	1	8	9	3
9	1	3	7	4	8	6	5	2
8	4	5	9	7	3	2	6	1
6	2	7	1	8	5	3	4	9
1	3	9	2	6	4	7	8	5

Sudoku # 144

8	2	6	5	3	4	9	1	7
1	7	3	8	2	9	4	6	5
5	4	9	1	6	7	8	2	3
3	9	8	2	5	1	6	7	4
7	6	2	4	8	3	1	5	9
4	1	5	9	7	6	2	3	8
9	5	7	6	1	8	3	4	2
6	3	4	7	9	2	5	8	1
2	8	1	3	4	5	7	9	6

Sudoku # 145

5	6	4	1	8	2	3	9	7
9	7	2	5	4	3	1	6	8
8	3	1	7	9	6	2	5	4
3	4	9	2	6	1	8	7	5
2	5	6	4	7	8	9	1	3
7	1	8	3	5	9	4	2	6
6	2	3	8	1	5	7	4	9
4	8	5	9	2	7	6	3	1
1	9	7	6	3	4	5	8	2

Sudoku # 146

9	3	7	8	4	5	1	6	2
6	8	1	9	3	2	5	7	4
4	2	5	6	7	1	9	8	3
1	4	3	2	9	6	8	5	7
7	6	2	4	5	8	3	1	9
8	5	9	7	1	3	4	2	6
5	1	6	3	2	9	7	4	8
3	7	8	1	6	4	2	9	5
2	9	4	5	8	7	6	3	1

Sudoku # 147

9	2	5	6	8	7	3	4	1
1	8	3	5	9	4	6	2	7
7	4	6	3	2	1	5	8	9
5	1	7	8	3	9	4	6	2
8	9	4	7	6	2	1	5	3
6	3	2	4	1	5	7	9	8
2	7	9	1	4	6	8	3	5
3	6	1	9	5	8	2	7	4
4	5	8	2	7	3	9	1	6

Sudoku # 148

3	9	4	2	8	5	7	1	6
7	5	1	6	3	9	2	4	8
8	6	2	7	1	4	9	3	5
4	7	6	8	9	3	1	5	2
9	2	3	5	6	1	8	7	4
5	1	8	4	2	7	3	6	9
1	3	5	9	4	2	6	8	7
6	4	9	3	7	8	5	2	1
2	8	7	1	5	6	4	9	3

Sudoku # 149

5	2	3	8	1	4	7	9	6
9	1	6	3	2	7	8	5	4
7	4	8	6	9	5	3	2	1
1	9	5	7	4	3	2	6	8
3	6	4	1	8	2	9	7	5
2	8	7	9	5	6	4	1	3
8	7	1	5	3	9	6	4	2
6	3	2	4	7	1	5	8	9
4	5	9	2	6	8	1	3	7

Sudoku # 150

1	3	4	7	5	9	2	6	8
9	6	2	4	1	8	5	7	3
5	7	8	3	2	6	1	9	4
8	4	3	5	6	7	9	2	1
7	5	1	9	8	2	3	4	6
6	2	9	1	3	4	8	5	7
4	1	6	2	9	3	7	8	5
3	9	7	8	4	5	6	1	2
2	8	5	6	7	1	4	3	9

Sudoku # 151

2	1	4	3	9	6	8	5	7
8	9	5	2	1	7	3	4	6
7	6	3	4	5	8	2	9	1
9	3	7	5	8	2	6	1	4
6	8	1	9	4	3	5	7	2
4	5	2	7	6	1	9	8	3
1	7	9	6	2	5	4	3	8
5	2	8	1	3	4	7	6	9
3	4	6	8	7	9	1	2	5

Sudoku # 152

1	2	5	6	7	3	9	8	4
4	7	8	5	1	9	2	6	3
3	6	9	4	2	8	7	5	1
6	3	7	2	5	4	1	9	8
9	5	1	7	8	6	3	4	2
8	4	2	9	3	1	5	7	6
5	9	6	1	4	2	8	3	7
2	8	4	3	9	7	6	1	5
7	1	3	8	6	5	4	2	9

Sudoku # 153

7	4	9	5	6	2	1	8	3
1	5	2	8	3	9	6	7	4
3	6	8	7	1	4	9	5	2
4	2	6	3	8	7	5	1	9
5	8	7	9	4	1	2	3	6
9	3	1	6	2	5	7	4	8
6	1	3	2	5	8	4	9	7
2	7	4	1	9	3	8	6	5
8	9	5	4	7	6	3	2	1

Sudoku # 154

4	2	7	8	1	6	5	9	3
1	5	3	2	9	7	8	4	6
9	6	8	5	4	3	7	1	2
6	4	1	7	8	2	3	5	9
5	8	2	6	3	9	4	7	1
3	7	9	4	5	1	6	2	8
7	3	4	1	2	8	9	6	5
2	9	6	3	7	5	1	8	4
8	1	5	9	6	4	2	3	7

Sudoku # 155

1	3	5	2	7	4	8	6	9
7	4	6	5	8	9	3	1	2
2	9	8	3	1	6	4	7	5
8	2	7	1	6	3	5	9	4
4	6	1	9	2	5	7	8	3
3	5	9	7	4	8	6	2	1
9	7	4	6	5	1	2	3	8
6	8	3	4	9	2	1	5	7
5	1	2	8	3	7	9	4	6

Sudoku # 156

5	4	3	7	8	9	2	1	6
6	2	8	3	4	1	5	7	9
7	1	9	5	6	2	4	8	3
1	7	2	9	3	4	6	5	8
8	6	4	1	7	5	3	9	2
9	3	5	6	2	8	1	4	7
2	8	6	4	5	7	9	3	1
4	9	7	2	1	3	8	6	5
3	5	1	8	9	6	7	2	4

Sudoku # 157

4	2	8	7	9	6	5	1	3
3	1	5	4	2	8	6	9	7
7	6	9	5	1	3	4	8	2
6	3	2	9	5	1	7	4	8
8	7	1	2	6	4	3	5	9
5	9	4	3	8	7	2	6	1
9	8	7	6	3	5	1	2	4
1	4	6	8	7	2	9	3	5
2	5	3	1	4	9	8	7	6

Sudoku # 158

9	1	5	2	3	7	6	8	4
7	8	3	9	4	6	1	2	5
2	6	4	1	5	8	9	7	3
6	5	2	3	8	4	7	1	9
8	4	9	6	7	1	5	3	2
3	7	1	5	2	9	4	6	8
5	9	7	8	6	2	3	4	1
4	3	8	7	1	5	2	9	6
1	2	6	4	9	3	8	5	7

Sudoku # 159

6	1	5	4	2	7	8	3	9
9	4	7	6	8	3	1	2	5
2	3	8	5	1	9	6	7	4
8	6	3	7	4	2	9	5	1
1	2	9	3	6	5	4	8	7
7	5	4	1	9	8	3	6	2
5	9	1	8	7	6	2	4	3
3	8	2	9	5	4	7	1	6
4	7	6	2	3	1	5	9	8

Sudoku # 160

7	2	1	8	3	9	6	5	4
6	8	5	4	1	2	3	7	9
9	4	3	7	6	5	2	1	8
3	5	2	1	4	8	9	6	7
1	6	8	2	9	7	4	3	5
4	7	9	3	5	6	1	8	2
2	1	7	9	8	3	5	4	6
5	9	4	6	7	1	8	2	3
8	3	6	5	2	4	7	9	1

Sudoku # 161

5	2	6	8	3	9	4	7	1
7	1	3	2	5	4	8	6	9
4	9	8	6	7	1	3	5	2
8	5	4	1	6	3	9	2	7
2	6	7	5	9	8	1	4	3
9	3	1	4	2	7	5	8	6
3	8	2	7	1	5	6	9	4
1	7	5	9	4	6	2	3	8
6	4	9	3	8	2	7	1	5

Sudoku # 162

3	9	8	4	5	2	6	1	7
2	1	4	3	6	7	8	5	9
5	6	7	1	9	8	3	4	2
4	5	9	6	1	3	7	2	8
6	8	2	9	7	4	5	3	1
1	7	3	8	2	5	4	9	6
8	3	1	2	4	6	9	7	5
9	4	5	7	8	1	2	6	3
7	2	6	5	3	9	1	8	4

Sudoku # 163

6	4	8	3	7	2	5	1	9
5	3	1	4	6	9	2	7	8
9	2	7	8	5	1	4	6	3
8	6	5	1	2	7	9	3	4
2	1	9	6	3	4	8	5	7
3	7	4	9	8	5	1	2	6
7	8	2	5	4	6	3	9	1
1	5	3	7	9	8	6	4	2
4	9	6	2	1	3	7	8	5

Sudoku # 164

7	3	2	1	6	8	4	5	9
4	8	5	3	2	9	6	1	7
6	9	1	5	4	7	2	3	8
2	6	9	8	7	1	5	4	3
8	1	3	4	5	6	9	7	2
5	7	4	2	9	3	8	6	1
3	2	7	6	8	4	1	9	5
1	4	8	9	3	5	7	2	6
9	5	6	7	1	2	3	8	4

Sudoku # 165

4	6	8	7	2	5	1	3	9
2	5	7	9	1	3	6	8	4
3	9	1	8	6	4	7	2	5
5	1	9	6	4	2	8	7	3
6	7	4	1	3	8	5	9	2
8	2	3	5	9	7	4	1	6
7	8	6	3	5	9	2	4	1
9	4	5	2	7	1	3	6	8
1	3	2	4	8	6	9	5	7

Sudoku # 166

1	6	8	2	4	9	3	5	7
4	2	7	1	3	5	9	6	8
5	9	3	8	7	6	4	1	2
3	4	9	7	8	1	5	2	6
6	8	2	3	5	4	7	9	1
7	1	5	9	6	2	8	3	4
9	3	6	4	2	7	1	8	5
2	7	1	5	9	8	6	4	3
8	5	4	6	1	3	2	7	9

Sudoku # 167

6	1	5	7	2	4	8	9	3
9	2	4	8	5	3	6	7	1
3	8	7	6	9	1	5	2	4
7	5	6	3	1	9	4	8	2
4	3	2	5	8	7	1	6	9
1	9	8	2	4	6	7	3	5
2	6	9	1	7	5	3	4	8
8	7	1	4	3	2	9	5	6
5	4	3	9	6	8	2	1	7

Sudoku # 168

9	6	2	1	4	7	3	8	5
7	3	5	8	2	9	1	4	6
1	8	4	5	6	3	9	7	2
5	1	7	9	8	2	4	6	3
8	9	3	4	7	6	2	5	1
4	2	6	3	1	5	8	9	7
3	7	1	6	9	4	5	2	8
6	4	8	2	5	1	7	3	9
2	5	9	7	3	8	6	1	4

Sudoku # 169

6	8	1	5	2	7	3	9	4
7	2	9	1	3	4	6	5	8
4	3	5	9	6	8	7	2	1
8	1	7	3	4	5	9	6	2
2	4	6	8	7	9	1	3	5
5	9	3	2	1	6	8	4	7
9	6	4	7	5	1	2	8	3
3	7	8	4	9	2	5	1	6
1	5	2	6	8	3	4	7	9

Sudoku # 170

1	7	5	9	4	6	3	2	8
9	8	4	5	2	3	6	7	1
2	3	6	1	7	8	5	4	9
4	2	9	7	6	5	8	1	3
3	6	1	4	8	9	7	5	2
7	5	8	3	1	2	4	9	6
5	1	2	6	3	7	9	8	4
8	9	3	2	5	4	1	6	7
6	4	7	8	9	1	2	3	5

Sudoku # 171

1	3	2	5	4	8	9	6	7
6	4	5	2	7	9	3	1	8
8	7	9	3	1	6	4	5	2
7	5	1	9	6	4	2	8	3
2	9	8	1	5	3	6	7	4
3	6	4	7	8	2	5	9	1
5	1	3	4	9	7	8	2	6
9	2	6	8	3	1	7	4	5
4	8	7	6	2	5	1	3	9

Sudoku # 172

3	7	9	8	4	5	1	6	2
5	6	2	3	9	1	8	7	4
1	8	4	2	7	6	3	5	9
6	4	3	5	1	9	2	8	7
7	5	1	4	2	8	9	3	6
9	2	8	6	3	7	5	4	1
8	3	7	9	6	2	4	1	5
4	9	6	1	5	3	7	2	8
2	1	5	7	8	4	6	9	3

Sudoku # 173

5	6	7	3	4	8	1	2	9
1	8	3	9	5	2	7	6	4
9	2	4	1	6	7	8	3	5
7	3	5	8	9	1	6	4	2
6	1	9	5	2	4	3	7	8
8	4	2	6	7	3	5	9	1
3	7	1	2	8	9	4	5	6
4	9	6	7	1	5	2	8	3
2	5	8	4	3	6	9	1	7

Sudoku # 174

9	3	6	8	4	7	5	1	2
4	1	7	9	5	2	8	6	3
8	5	2	1	6	3	7	9	4
3	2	8	4	7	1	9	5	6
6	9	5	3	2	8	1	4	7
7	4	1	5	9	6	2	3	8
2	8	3	6	1	9	4	7	5
5	7	9	2	3	4	6	8	1
1	6	4	7	8	5	3	2	9

Sudoku # 175

3	9	8	4	2	6	5	7	1
4	7	1	8	5	9	6	3	2
6	5	2	7	1	3	4	9	8
5	2	7	6	4	8	9	1	3
9	6	4	2	3	1	7	8	5
8	1	3	5	9	7	2	6	4
7	3	5	1	6	2	8	4	9
1	4	6	9	8	5	3	2	7
2	8	9	3	7	4	1	5	6

Sudoku # 176

5	4	9	6	2	7	1	3	8
2	7	1	3	8	9	4	6	5
8	3	6	4	5	1	7	2	9
6	9	2	8	4	5	3	1	7
3	5	8	1	7	6	9	4	2
4	1	7	2	9	3	5	8	6
9	8	4	5	1	2	6	7	3
1	6	5	7	3	8	2	9	4
7	2	3	9	6	4	8	5	1

Sudoku # 177

5	7	2	8	1	9	4	3	6
1	4	8	5	6	3	2	9	7
9	3	6	7	4	2	1	8	5
7	2	5	9	3	6	8	1	4
8	9	1	2	5	4	7	6	3
3	6	4	1	7	8	9	5	2
6	5	7	4	8	1	3	2	9
4	8	9	3	2	5	6	7	1
2	1	3	6	9	7	5	4	8

Sudoku # 178

4	6	9	5	8	3	2	7	1
2	8	3	9	1	7	5	4	6
1	7	5	4	2	6	9	8	3
5	3	4	8	9	2	6	1	7
6	2	7	1	3	5	4	9	8
9	1	8	6	7	4	3	2	5
8	9	6	2	5	1	7	3	4
7	5	2	3	4	8	1	6	9
3	4	1	7	6	9	8	5	2

Sudoku # 179

4	6	7	8	1	3	2	9	5
2	5	8	7	4	9	1	6	3
1	3	9	2	6	5	7	8	4
8	2	5	9	7	6	3	4	1
3	7	1	4	5	8	6	2	9
6	9	4	1	3	2	5	7	8
9	1	2	3	8	7	4	5	6
7	4	6	5	9	1	8	3	2
5	8	3	6	2	4	9	1	7

Sudoku # 180

9	7	6	8	4	1	2	3	5
2	4	3	5	6	9	1	8	7
5	8	1	2	3	7	4	6	9
6	1	5	7	8	3	9	2	4
3	2	8	4	9	5	6	7	1
4	9	7	6	1	2	3	5	8
7	5	4	9	2	6	8	1	3
8	3	2	1	5	4	7	9	6
1	6	9	3	7	8	5	4	2

Sudoku # 181

4	3	5	1	8	9	2	7	6
1	9	7	5	2	6	8	4	3
2	6	8	7	3	4	5	1	9
8	2	3	9	1	5	7	6	4
6	1	4	8	7	2	9	3	5
5	7	9	6	4	3	1	8	2
3	8	2	4	9	7	6	5	1
7	4	6	2	5	1	3	9	8
9	5	1	3	6	8	4	2	7

Sudoku # 182

9	1	5	6	2	8	4	7	3
7	3	4	5	1	9	2	6	8
2	6	8	7	4	3	9	5	1
5	4	1	9	3	7	8	2	6
6	8	9	1	5	2	3	4	7
3	7	2	8	6	4	5	1	9
8	5	6	2	9	1	7	3	4
1	9	3	4	7	5	6	8	2
4	2	7	3	8	6	1	9	5

Sudoku # 183

3	8	4	9	7	5	1	6	2
5	2	1	3	4	6	8	9	7
6	7	9	1	2	8	3	4	5
1	3	7	4	5	2	6	8	9
4	6	2	8	1	9	7	5	3
8	9	5	7	6	3	2	1	4
7	5	8	6	3	4	9	2	1
9	4	3	2	8	1	5	7	6
2	1	6	5	9	7	4	3	8

Sudoku # 184

6	3	8	9	5	4	7	1	2
2	7	5	6	3	1	8	4	9
9	4	1	2	8	7	6	5	3
3	5	4	7	2	6	9	8	1
7	6	9	3	1	8	4	2	5
1	8	2	5	4	9	3	6	7
8	9	6	1	7	5	2	3	4
5	2	7	4	6	3	1	9	8
4	1	3	8	9	2	5	7	6

Sudoku # 185

4	9	6	2	7	1	5	8	3
3	5	7	8	4	6	9	2	1
2	1	8	3	5	9	4	6	7
8	3	1	4	9	5	6	7	2
5	6	2	7	8	3	1	9	4
7	4	9	1	6	2	3	5	8
1	7	5	9	3	8	2	4	6
6	8	3	5	2	4	7	1	9
9	2	4	6	1	7	8	3	5

Sudoku # 186

1	2	5	3	8	4	9	6	7
6	7	4	1	9	5	8	2	3
9	3	8	7	2	6	4	5	1
7	8	3	2	5	9	6	1	4
2	6	9	8	4	1	7	3	5
4	5	1	6	3	7	2	9	8
3	1	6	9	7	8	5	4	2
8	4	2	5	6	3	1	7	9
5	9	7	4	1	2	3	8	6

Sudoku # 187

5	1	4	6	9	8	3	2	7
3	9	7	5	2	4	8	1	6
2	8	6	3	7	1	5	4	9
8	2	9	7	4	5	1	6	3
1	6	5	8	3	9	2	7	4
7	4	3	2	1	6	9	5	8
6	7	1	9	8	2	4	3	5
4	3	8	1	5	7	6	9	2
9	5	2	4	6	3	7	8	1

Sudoku # 188

4	5	3	6	7	9	1	8	2
2	1	9	4	3	8	5	6	7
7	6	8	2	5	1	3	4	9
1	7	5	9	2	4	6	3	8
3	2	6	8	1	5	7	9	4
9	8	4	7	6	3	2	5	1
8	3	1	5	9	2	4	7	6
5	9	7	1	4	6	8	2	3
6	4	2	3	8	7	9	1	5

Sudoku # 189

2	8	3	1	4	7	6	9	5
9	4	6	8	5	3	1	7	2
5	1	7	2	9	6	3	4	8
3	9	4	5	8	2	7	6	1
8	7	1	9	6	4	2	5	3
6	5	2	3	7	1	4	8	9
4	3	9	7	2	8	5	1	6
1	6	5	4	3	9	8	2	7
7	2	8	6	1	5	9	3	4

Sudoku # 190

7	8	1	3	9	5	2	6	4
2	4	9	1	6	7	3	5	8
3	6	5	8	2	4	9	7	1
5	9	8	6	4	1	7	3	2
6	1	3	5	7	2	8	4	9
4	7	2	9	3	8	6	1	5
8	2	4	7	5	3	1	9	6
1	3	6	4	8	9	5	2	7
9	5	7	2	1	6	4	8	3

Sudoku # 191

6	2	5	9	3	1	8	7	4
7	3	9	5	4	8	1	6	2
8	1	4	6	7	2	3	5	9
5	8	7	3	9	4	2	1	6
1	9	2	7	8	6	5	4	3
3	4	6	1	2	5	9	8	7
4	7	8	2	5	9	6	3	1
2	6	3	8	1	7	4	9	5
9	5	1	4	6	3	7	2	8

Sudoku # 192

9	5	6	1	2	3	8	7	4
8	4	3	7	5	9	2	1	6
1	7	2	8	4	6	5	9	3
5	8	7	3	1	2	6	4	9
2	9	1	5	6	4	7	3	8
3	6	4	9	7	8	1	2	5
7	1	8	4	3	5	9	6	2
4	2	9	6	8	1	3	5	7
6	3	5	2	9	7	4	8	1

Sudoku # 193

7	3	9	6	1	8	4	5	2
2	5	6	9	3	4	1	7	8
1	4	8	5	2	7	6	9	3
9	6	7	8	4	3	5	2	1
4	2	1	7	9	5	8	3	6
5	8	3	2	6	1	7	4	9
6	7	5	3	8	9	2	1	4
3	1	2	4	5	6	9	8	7
8	9	4	1	7	2	3	6	5

Sudoku # 194

4	3	2	1	5	9	7	8	6
5	6	1	4	8	7	2	9	3
7	9	8	3	6	2	1	5	4
6	4	3	9	7	5	8	2	1
1	7	9	2	3	8	6	4	5
2	8	5	6	1	4	9	3	7
8	2	7	5	4	1	3	6	9
3	1	4	8	9	6	5	7	2
9	5	6	7	2	3	4	1	8

Sudoku # 195

6	8	2	3	4	5	7	9	1
1	9	7	8	6	2	4	3	5
5	3	4	9	7	1	8	6	2
4	6	5	7	9	8	2	1	3
7	2	3	4	1	6	9	5	8
8	1	9	2	5	3	6	4	7
2	7	1	6	3	4	5	8	9
9	5	6	1	8	7	3	2	4
3	4	8	5	2	9	1	7	6

Sudoku # 196

3	5	7	6	4	8	1	9	2
6	4	1	3	2	9	8	5	7
8	2	9	7	5	1	6	3	4
7	6	2	9	3	5	4	1	8
4	3	8	2	1	7	5	6	9
9	1	5	4	8	6	2	7	3
5	9	4	1	7	2	3	8	6
1	7	3	8	6	4	9	2	5
2	8	6	5	9	3	7	4	1

Sudoku # 197

5	6	2	8	4	9	3	7	1
4	3	8	7	1	5	2	6	9
1	9	7	2	6	3	5	8	4
8	4	6	5	2	7	1	9	3
9	5	3	6	8	1	7	4	2
7	2	1	3	9	4	8	5	6
6	1	5	4	3	8	9	2	7
3	7	4	9	5	2	6	1	8
2	8	9	1	7	6	4	3	5

Sudoku # 198

6	8	1	9	5	7	3	2	4
3	4	2	6	8	1	9	7	5
5	9	7	3	4	2	1	8	6
4	3	9	5	6	8	7	1	2
7	6	5	1	2	9	8	4	3
1	2	8	7	3	4	5	6	9
8	7	4	2	9	3	6	5	1
9	1	6	4	7	5	2	3	8
2	5	3	8	1	6	4	9	7

Sudoku # 199

9	3	4	1	6	2	8	5	7
7	8	1	5	9	4	3	2	6
2	6	5	8	7	3	4	9	1
5	9	6	7	4	1	2	3	8
1	2	3	6	5	8	9	7	4
4	7	8	3	2	9	1	6	5
8	1	7	9	3	6	5	4	2
3	5	2	4	8	7	6	1	9
6	4	9	2	1	5	7	8	3

Sudoku # 200

8	5	1	9	6	3	7	2	4
4	2	7	5	8	1	6	9	3
3	6	9	2	4	7	8	5	1
6	7	4	1	5	8	2	3	9
1	9	3	4	2	6	5	8	7
5	8	2	3	7	9	1	4	6
9	1	5	7	3	2	4	6	8
7	4	6	8	9	5	3	1	2
2	3	8	6	1	4	9	7	5

Sudoku # 201

9	8	5	3	1	7	4	2	6
4	2	7	5	6	9	3	1	8
3	6	1	2	8	4	7	5	9
2	3	8	6	7	5	1	9	4
7	5	4	1	9	8	6	3	2
6	1	9	4	2	3	5	8	7
8	4	6	9	3	1	2	7	5
1	9	2	7	5	6	8	4	3
5	7	3	8	4	2	9	6	1

Sudoku # 202

5	3	4	1	7	2	8	6	9
7	1	8	4	6	9	2	5	3
9	6	2	5	3	8	4	7	1
2	9	3	7	5	6	1	4	8
1	5	7	2	8	4	9	3	6
4	8	6	3	9	1	7	2	5
3	7	9	8	4	5	6	1	2
6	2	5	9	1	7	3	8	4
8	4	1	6	2	3	5	9	7

Sudoku # 203

2	8	1	6	5	9	3	4	7
6	4	9	7	1	3	2	8	5
7	3	5	4	8	2	1	9	6
9	1	3	5	6	8	7	2	4
8	7	6	2	4	1	9	5	3
4	5	2	9	3	7	6	1	8
1	9	4	3	7	5	8	6	2
5	2	7	8	9	6	4	3	1
3	6	8	1	2	4	5	7	9

Sudoku # 204

1	3	5	9	2	8	4	7	6
4	8	9	7	1	6	3	5	2
7	2	6	4	3	5	9	8	1
6	4	2	1	7	9	8	3	5
3	1	7	5	8	4	2	6	9
9	5	8	3	6	2	1	4	7
5	6	3	8	9	1	7	2	4
2	7	1	6	4	3	5	9	8
8	9	4	2	5	7	6	1	3

Sudoku # 205

3	7	2	5	8	4	6	9	1
6	1	4	2	7	9	5	8	3
9	8	5	6	3	1	7	2	4
5	4	9	1	6	2	3	7	8
7	6	1	8	5	3	2	4	9
8	2	3	9	4	7	1	6	5
1	3	8	7	9	6	4	5	2
2	9	7	4	1	5	8	3	6
4	5	6	3	2	8	9	1	7

Sudoku # 206

5	4	9	3	8	6	1	2	7
3	8	6	7	1	2	5	9	4
2	1	7	5	9	4	6	3	8
1	3	5	6	7	8	2	4	9
4	7	2	1	5	9	8	6	3
9	6	8	4	2	3	7	1	5
6	5	1	9	4	7	3	8	2
7	2	4	8	3	1	9	5	6
8	9	3	2	6	5	4	7	1

Sudoku # 207

4	7	1	6	8	9	3	2	5
2	5	9	3	1	4	6	8	7
3	8	6	2	5	7	9	4	1
8	3	2	1	9	6	5	7	4
7	6	4	5	3	2	1	9	8
9	1	5	4	7	8	2	6	3
5	2	7	9	4	3	8	1	6
1	9	8	7	6	5	4	3	2
6	4	3	8	2	1	7	5	9

Sudoku # 208

6	9	2	4	8	3	1	7	5
5	3	8	1	9	7	2	6	4
4	7	1	5	6	2	8	9	3
3	6	5	2	4	8	7	1	9
8	1	4	9	7	5	6	3	2
9	2	7	6	3	1	4	5	8
7	8	9	3	1	4	5	2	6
2	4	6	7	5	9	3	8	1
1	5	3	8	2	6	9	4	7

Sudoku # 209

6	7	8	4	5	2	1	9	3
9	4	5	7	3	1	2	6	8
3	1	2	6	9	8	5	7	4
1	6	9	3	4	7	8	2	5
7	8	4	9	2	5	6	3	1
5	2	3	8	1	6	7	4	9
2	9	7	5	8	3	4	1	6
8	3	1	2	6	4	9	5	7
4	5	6	1	7	9	3	8	2

Sudoku # 210

5	1	6	8	4	7	2	3	9
7	9	3	5	1	2	8	6	4
8	4	2	3	6	9	5	1	7
2	8	5	7	9	6	3	4	1
6	3	1	2	5	4	9	7	8
4	7	9	1	8	3	6	2	5
9	6	8	4	3	1	7	5	2
3	2	4	9	7	5	1	8	6
1	5	7	6	2	8	4	9	3

Sudoku # 211

9	1	8	5	7	6	4	3	2
2	3	7	9	1	4	8	5	6
5	4	6	8	3	2	1	7	9
1	9	2	7	6	3	5	4	8
8	6	5	1	4	9	3	2	7
4	7	3	2	8	5	9	6	1
7	8	4	3	2	1	6	9	5
3	2	9	6	5	8	7	1	4
6	5	1	4	9	7	2	8	3

Sudoku # 212

3	5	1	6	7	4	2	9	8
6	7	2	1	9	8	5	3	4
4	8	9	5	3	2	1	6	7
8	3	4	2	6	5	9	7	1
5	2	7	8	1	9	3	4	6
1	9	6	3	4	7	8	2	5
9	6	3	4	8	1	7	5	2
2	4	8	7	5	3	6	1	9
7	1	5	9	2	6	4	8	3

Sudoku # 213

2	4	9	5	1	8	6	3	7
5	1	8	7	6	3	2	9	4
6	3	7	9	4	2	5	8	1
4	9	5	3	7	6	1	2	8
1	7	2	4	8	5	9	6	3
3	8	6	2	9	1	4	7	5
8	6	4	1	3	9	7	5	2
9	2	1	8	5	7	3	4	6
7	5	3	6	2	4	8	1	9

Sudoku # 214

6	1	4	8	3	7	9	5	2
5	7	8	1	9	2	6	3	4
9	3	2	6	4	5	8	1	7
3	4	9	5	2	1	7	6	8
7	2	1	3	8	6	4	9	5
8	6	5	4	7	9	1	2	3
2	5	6	7	1	8	3	4	9
1	8	3	9	5	4	2	7	6
4	9	7	2	6	3	5	8	1

Sudoku # 215

9	6	7	8	2	5	3	1	4
8	2	4	6	3	1	5	7	9
5	3	1	9	7	4	6	2	8
3	7	9	1	4	8	2	6	5
1	5	8	7	6	2	4	9	3
6	4	2	5	9	3	7	8	1
2	8	6	4	5	9	1	3	7
7	9	5	3	1	6	8	4	2
4	1	3	2	8	7	9	5	6

Sudoku # 216

7	8	5	9	2	6	4	1	3
1	4	6	3	7	5	2	8	9
9	3	2	4	8	1	6	7	5
6	2	3	1	4	8	5	9	7
8	5	9	6	3	7	1	2	4
4	7	1	2	5	9	3	6	8
5	6	7	8	1	4	9	3	2
2	1	8	5	9	3	7	4	6
3	9	4	7	6	2	8	5	1

Sudoku # 217

6	1	3	8	7	4	5	9	2
2	7	5	9	6	1	8	3	4
9	8	4	2	5	3	6	1	7
5	6	2	7	1	9	4	8	3
1	3	9	4	2	8	7	6	5
7	4	8	6	3	5	9	2	1
3	5	6	1	9	7	2	4	8
4	9	7	3	8	2	1	5	6
8	2	1	5	4	6	3	7	9

Sudoku # 218

6	7	5	4	2	1	3	9	8
1	3	8	7	6	9	2	5	4
2	4	9	5	8	3	1	7	6
3	2	1	6	5	8	9	4	7
5	8	7	2	9	4	6	3	1
4	9	6	3	1	7	5	8	2
9	6	3	8	4	2	7	1	5
7	5	4	1	3	6	8	2	9
8	1	2	9	7	5	4	6	3

Sudoku # 219

1	7	4	6	3	9	8	2	5
5	3	9	8	2	1	7	6	4
2	6	8	5	4	7	9	1	3
9	8	2	3	7	5	6	4	1
6	1	3	4	8	2	5	7	9
4	5	7	9	1	6	3	8	2
7	2	6	1	9	3	4	5	8
3	4	1	7	5	8	2	9	6
8	9	5	2	6	4	1	3	7

Sudoku # 220

1	7	3	2	5	6	4	8	9
8	2	9	1	7	4	6	5	3
4	6	5	8	3	9	1	2	7
5	1	6	7	2	3	8	9	4
7	4	8	6	9	5	3	1	2
9	3	2	4	8	1	5	7	6
2	9	4	3	1	8	7	6	5
3	8	7	5	6	2	9	4	1
6	5	1	9	4	7	2	3	8

Sudoku # 221

4	8	1	2	6	7	9	5	3
5	2	3	4	1	9	8	7	6
7	6	9	3	8	5	1	4	2
6	5	4	1	2	8	3	9	7
9	7	8	6	5	3	2	1	4
3	1	2	7	9	4	5	6	8
2	3	6	5	4	1	7	8	9
8	4	5	9	7	2	6	3	1
1	9	7	8	3	6	4	2	5

Sudoku # 222

6	5	8	9	3	1	4	2	7
9	2	7	6	4	8	3	5	1
4	1	3	5	2	7	9	8	6
3	6	2	4	9	5	1	7	8
8	7	5	3	1	2	6	4	9
1	4	9	7	8	6	2	3	5
7	3	1	2	5	9	8	6	4
2	8	6	1	7	4	5	9	3
5	9	4	8	6	3	7	1	2

Sudoku # 223

8	1	3	4	7	2	6	5	9
2	9	4	8	5	6	1	7	3
6	7	5	9	3	1	8	2	4
3	6	2	7	9	4	5	1	8
5	8	7	1	2	3	4	9	6
9	4	1	5	6	8	7	3	2
7	5	6	3	4	9	2	8	1
4	3	8	2	1	5	9	6	7
1	2	9	6	8	7	3	4	5

Sudoku # 224

9	3	1	8	5	7	6	2	4
5	8	2	3	4	6	9	1	7
4	7	6	9	2	1	8	3	5
7	9	8	2	6	4	1	5	3
6	4	5	7	1	3	2	9	8
2	1	3	5	9	8	7	4	6
8	5	9	6	3	2	4	7	1
3	6	4	1	7	9	5	8	2
1	2	7	4	8	5	3	6	9

Sudoku # 225

3	8	5	9	7	4	6	1	2
1	6	2	8	3	5	7	4	9
7	9	4	1	6	2	3	8	5
8	3	1	5	9	7	2	6	4
6	4	7	2	8	3	5	9	1
5	2	9	4	1	6	8	3	7
2	1	3	6	5	9	4	7	8
9	5	6	7	4	8	1	2	3
4	7	8	3	2	1	9	5	6

Sudoku # 226

3	5	9	8	1	7	4	2	6
6	8	1	4	5	2	9	3	7
4	2	7	9	3	6	8	1	5
7	6	3	2	8	5	1	9	4
8	9	5	7	4	1	3	6	2
2	1	4	3	6	9	7	5	8
1	4	8	6	2	3	5	7	9
9	3	2	5	7	8	6	4	1
5	7	6	1	9	4	2	8	3

Sudoku # 227

6	9	5	4	7	3	2	1	8
8	2	4	9	6	1	3	5	7
7	3	1	2	8	5	4	9	6
9	1	3	7	5	2	8	6	4
5	6	7	3	4	8	9	2	1
4	8	2	6	1	9	7	3	5
3	4	6	5	9	7	1	8	2
2	7	8	1	3	6	5	4	9
1	5	9	8	2	4	6	7	3

Sudoku # 228

1	6	4	5	7	9	8	2	3
2	5	3	8	1	4	9	6	7
7	9	8	2	6	3	1	5	4
5	2	7	1	3	8	4	9	6
3	8	1	4	9	6	5	7	2
6	4	9	7	5	2	3	1	8
9	7	2	3	4	5	6	8	1
4	1	6	9	8	7	2	3	5
8	3	5	6	2	1	7	4	9

Sudoku # 229

8	1	3	7	4	2	5	6	9
6	2	4	8	9	5	1	3	7
5	7	9	3	6	1	8	2	4
3	4	5	2	1	9	7	8	6
2	8	6	4	3	7	9	1	5
7	9	1	6	5	8	2	4	3
1	3	2	5	7	6	4	9	8
9	6	7	1	8	4	3	5	2
4	5	8	9	2	3	6	7	1

Sudoku # 230

6	7	2	4	9	8	5	1	3
1	9	4	5	6	3	2	8	7
3	8	5	7	1	2	6	9	4
9	3	8	2	5	7	4	6	1
7	5	6	3	4	1	8	2	9
2	4	1	9	8	6	7	3	5
4	1	3	8	2	5	9	7	6
8	6	9	1	7	4	3	5	2
5	2	7	6	3	9	1	4	8

Sudoku # 231

8	6	5	1	4	2	9	7	3
1	4	2	7	9	3	5	6	8
9	7	3	8	6	5	1	2	4
3	8	1	9	5	6	7	4	2
6	2	4	3	1	7	8	5	9
7	5	9	2	8	4	3	1	6
5	9	8	6	2	1	4	3	7
2	1	7	4	3	8	6	9	5
4	3	6	5	7	9	2	8	1

Sudoku # 232

7	3	1	2	8	9	6	5	4
2	6	8	7	4	5	1	3	9
5	4	9	6	1	3	8	2	7
8	9	2	3	5	4	7	6	1
3	7	5	8	6	1	4	9	2
6	1	4	9	7	2	3	8	5
9	5	7	1	3	6	2	4	8
4	8	3	5	2	7	9	1	6
1	2	6	4	9	8	5	7	3

Sudoku # 233

9	3	1	7	5	2	4	8	6
7	6	8	1	9	4	2	3	5
5	2	4	6	8	3	1	7	9
4	9	7	5	3	8	6	1	2
8	1	6	2	4	7	5	9	3
3	5	2	9	6	1	8	4	7
2	8	9	3	1	5	7	6	4
6	4	5	8	7	9	3	2	1
1	7	3	4	2	6	9	5	8

Sudoku # 234

1	8	2	7	6	3	9	5	4
4	3	5	2	1	9	8	7	6
9	7	6	8	4	5	3	1	2
8	1	9	4	2	7	6	3	5
2	6	3	5	9	1	4	8	7
5	4	7	3	8	6	2	9	1
6	2	1	9	7	8	5	4	3
7	5	8	6	3	4	1	2	9
3	9	4	1	5	2	7	6	8

Sudoku # 235

8	6	5	2	3	7	9	4	1
1	3	9	4	8	6	7	5	2
2	7	4	5	1	9	3	8	6
4	5	7	1	9	3	2	6	8
9	8	2	6	5	4	1	3	7
3	1	6	7	2	8	5	9	4
7	4	3	9	6	2	8	1	5
5	2	8	3	4	1	6	7	9
6	9	1	8	7	5	4	2	3

Sudoku # 236

6	5	2	8	1	4	7	9	3
4	8	1	7	3	9	5	6	2
9	3	7	2	6	5	8	4	1
5	1	9	4	7	3	2	8	6
3	4	6	5	8	2	9	1	7
2	7	8	6	9	1	3	5	4
8	6	5	3	4	7	1	2	9
1	2	3	9	5	6	4	7	8
7	9	4	1	2	8	6	3	5

Sudoku # 237

1	9	3	7	5	6	4	2	8
2	7	4	1	9	8	5	3	6
8	5	6	4	3	2	9	7	1
4	3	2	5	6	9	1	8	7
9	1	8	3	2	7	6	4	5
5	6	7	8	4	1	3	9	2
6	8	1	9	7	4	2	5	3
7	4	5	2	1	3	8	6	9
3	2	9	6	8	5	7	1	4

Sudoku # 238

4	3	2	8	6	9	1	7	5
5	8	9	2	7	1	4	3	6
7	6	1	4	3	5	8	2	9
1	9	5	3	8	4	2	6	7
8	7	4	5	2	6	3	9	1
3	2	6	1	9	7	5	4	8
9	4	8	7	5	2	6	1	3
6	1	3	9	4	8	7	5	2
2	5	7	6	1	3	9	8	4

Sudoku # 239

9	3	4	2	5	7	8	1	6
5	7	1	4	6	8	3	2	9
2	6	8	1	9	3	5	7	4
6	1	7	5	3	2	4	9	8
8	2	3	7	4	9	6	5	1
4	5	9	6	8	1	2	3	7
1	4	5	9	2	6	7	8	3
7	8	2	3	1	4	9	6	5
3	9	6	8	7	5	1	4	2

Sudoku # 240

3	4	5	6	7	1	9	2	8
9	1	2	4	5	8	3	7	6
8	6	7	2	9	3	4	5	1
5	2	8	3	1	7	6	4	9
1	7	9	8	6	4	5	3	2
4	3	6	5	2	9	1	8	7
7	8	1	9	3	5	2	6	4
2	9	3	7	4	6	8	1	5
6	5	4	1	8	2	7	9	3

Sudoku # 241

6	4	3	2	5	7	8	9	1
9	1	8	6	4	3	2	7	5
5	7	2	8	9	1	6	4	3
3	5	1	4	2	6	7	8	9
2	9	6	1	7	8	5	3	4
4	8	7	5	3	9	1	2	6
8	6	4	3	1	2	9	5	7
1	3	9	7	8	5	4	6	2
7	2	5	9	6	4	3	1	8

Sudoku # 242

9	3	8	1	5	2	4	7	6
2	7	5	9	6	4	1	3	8
6	1	4	7	8	3	2	5	9
3	6	1	8	2	5	7	9	4
8	4	7	3	9	1	5	6	2
5	2	9	4	7	6	3	8	1
7	9	2	5	4	8	6	1	3
4	5	3	6	1	9	8	2	7
1	8	6	2	3	7	9	4	5

Sudoku # 243

6	5	4	9	8	1	7	3	2
9	8	1	2	3	7	4	6	5
2	3	7	5	4	6	1	8	9
4	6	8	7	5	2	9	1	3
5	9	3	4	1	8	6	2	7
1	7	2	3	6	9	8	5	4
3	4	6	8	9	5	2	7	1
7	1	9	6	2	3	5	4	8
8	2	5	1	7	4	3	9	6

Sudoku # 244

2	7	6	9	5	4	8	1	3
3	1	5	6	2	8	7	4	9
4	8	9	3	7	1	5	2	6
1	5	4	8	6	9	2	3	7
7	9	3	4	1	2	6	5	8
6	2	8	5	3	7	4	9	1
9	4	1	2	8	6	3	7	5
5	6	2	7	9	3	1	8	4
8	3	7	1	4	5	9	6	2

Sudoku # 245

2	7	3	9	1	6	4	8	5
6	9	8	5	2	4	3	1	7
5	4	1	8	3	7	6	2	9
8	5	6	7	9	3	2	4	1
9	1	2	6	4	5	7	3	8
4	3	7	2	8	1	5	9	6
3	2	5	1	6	9	8	7	4
1	6	4	3	7	8	9	5	2
7	8	9	4	5	2	1	6	3

Sudoku # 246

2	1	7	9	4	3	5	6	8
4	5	6	7	8	1	2	9	3
3	8	9	2	5	6	1	7	4
7	4	8	6	1	5	3	2	9
1	6	2	4	3	9	7	8	5
5	9	3	8	2	7	6	4	1
8	3	4	5	6	2	9	1	7
9	2	1	3	7	8	4	5	6
6	7	5	1	9	4	8	3	2

Sudoku # 247

4	6	2	7	3	1	5	9	8
8	3	7	6	5	9	2	1	4
9	5	1	2	4	8	7	6	3
2	4	9	1	6	7	8	3	5
7	8	3	9	2	5	1	4	6
6	1	5	4	8	3	9	7	2
1	2	8	3	7	6	4	5	9
5	9	6	8	1	4	3	2	7
3	7	4	5	9	2	6	8	1

Sudoku # 248

7	3	2	6	4	1	8	9	5
1	4	6	8	5	9	2	3	7
9	5	8	7	3	2	6	4	1
3	7	4	9	1	8	5	2	6
8	2	9	3	6	5	7	1	4
5	6	1	4	2	7	9	8	3
2	1	3	5	9	6	4	7	8
4	8	5	2	7	3	1	6	9
6	9	7	1	8	4	3	5	2

Sudoku # 249

6	3	1	7	2	8	5	9	4
4	8	7	5	3	9	1	2	6
9	5	2	4	6	1	7	3	8
5	4	9	8	1	2	6	7	3
8	1	6	3	5	7	9	4	2
2	7	3	6	9	4	8	1	5
7	6	8	1	4	3	2	5	9
1	9	4	2	8	5	3	6	7
3	2	5	9	7	6	4	8	1

Sudoku # 250

2	4	8	3	1	6	9	5	7
3	7	6	4	9	5	1	8	2
1	9	5	2	7	8	6	3	4
6	8	3	7	4	1	5	2	9
5	1	7	6	2	9	8	4	3
4	2	9	5	8	3	7	1	6
7	6	4	8	5	2	3	9	1
9	5	2	1	3	7	4	6	8
8	3	1	9	6	4	2	7	5

Sudoku # 251

1	8	7	6	3	9	5	2	4
4	5	6	7	2	1	9	3	8
9	2	3	5	8	4	7	1	6
3	9	5	2	1	6	4	8	7
2	7	4	8	5	3	6	9	1
6	1	8	9	4	7	2	5	3
7	4	2	3	9	8	1	6	5
8	6	9	1	7	5	3	4	2
5	3	1	4	6	2	8	7	9

Sudoku # 252

8	6	4	9	3	2	7	1	5
3	5	2	1	4	7	6	9	8
1	7	9	8	6	5	4	2	3
2	8	3	4	5	1	9	6	7
9	1	7	2	8	6	5	3	4
5	4	6	3	7	9	1	8	2
4	3	5	6	9	8	2	7	1
6	2	8	7	1	4	3	5	9
7	9	1	5	2	3	8	4	6

Sudoku # 253

2	7	6	8	4	5	3	9	1
8	3	9	6	1	7	5	4	2
5	4	1	3	9	2	7	8	6
9	6	2	1	7	3	4	5	8
7	1	4	9	5	8	6	2	3
3	5	8	4	2	6	9	1	7
6	2	5	7	8	9	1	3	4
4	9	3	2	6	1	8	7	5
1	8	7	5	3	4	2	6	9

Sudoku # 254

1	9	8	2	7	6	4	3	5
5	7	3	1	8	4	6	2	9
4	6	2	5	3	9	1	8	7
9	3	1	6	5	2	8	7	4
8	4	7	9	1	3	5	6	2
6	2	5	8	4	7	9	1	3
7	8	9	4	2	1	3	5	6
2	5	4	3	6	8	7	9	1
3	1	6	7	9	5	2	4	8

Sudoku # 255

1	3	9	8	4	2	6	5	7
8	7	4	1	5	6	2	9	3
5	6	2	3	9	7	8	1	4
3	1	7	6	8	9	4	2	5
9	2	5	4	1	3	7	8	6
6	4	8	2	7	5	1	3	9
2	5	1	7	3	4	9	6	8
4	8	3	9	6	1	5	7	2
7	9	6	5	2	8	3	4	1

Sudoku # 256

6	1	4	3	5	2	8	9	7
3	5	9	7	4	8	1	2	6
8	2	7	9	6	1	4	5	3
2	6	8	4	7	3	9	1	5
7	3	1	2	9	5	6	8	4
4	9	5	8	1	6	3	7	2
5	7	3	1	8	4	2	6	9
9	8	2	6	3	7	5	4	1
1	4	6	5	2	9	7	3	8

Sudoku # 257

1	4	7	8	3	2	6	9	5
6	3	5	4	9	7	1	2	8
9	2	8	5	6	1	7	4	3
2	5	4	7	8	3	9	6	1
3	9	6	1	4	5	2	8	7
7	8	1	6	2	9	3	5	4
4	1	3	9	5	6	8	7	2
5	6	2	3	7	8	4	1	9
8	7	9	2	1	4	5	3	6

Sudoku # 258

5	7	9	2	3	8	4	6	1
4	8	1	6	9	5	3	2	7
2	3	6	1	7	4	9	8	5
1	4	7	9	8	3	2	5	6
8	9	3	5	2	6	1	7	4
6	2	5	7	4	1	8	3	9
7	5	2	8	1	9	6	4	3
3	1	8	4	6	7	5	9	2
9	6	4	3	5	2	7	1	8

Sudoku # 259

4	1	9	7	2	5	8	6	3
7	5	6	9	3	8	2	1	4
2	3	8	1	4	6	9	7	5
6	7	2	8	1	3	4	5	9
1	9	4	5	6	2	7	3	8
3	8	5	4	9	7	6	2	1
9	6	1	2	5	4	3	8	7
8	4	3	6	7	1	5	9	2
5	2	7	3	8	9	1	4	6

Sudoku # 260

9	1	4	5	2	8	3	7	6
5	6	2	7	1	3	9	8	4
3	8	7	4	9	6	1	5	2
4	7	9	8	3	2	6	1	5
1	3	5	9	6	7	2	4	8
6	2	8	1	4	5	7	9	3
7	4	6	2	5	9	8	3	1
8	5	3	6	7	1	4	2	9
2	9	1	3	8	4	5	6	7

Sudoku # 261

9	2	7	3	5	8	4	1	6
3	8	4	2	1	6	7	9	5
6	1	5	7	4	9	2	3	8
7	9	6	5	8	1	3	4	2
1	3	2	6	7	4	5	8	9
4	5	8	9	3	2	6	7	1
8	4	3	1	2	5	9	6	7
5	6	1	4	9	7	8	2	3
2	7	9	8	6	3	1	5	4

Sudoku # 262

4	7	5	6	3	8	2	9	1
8	6	2	5	1	9	7	3	4
9	3	1	7	4	2	6	5	8
6	8	7	4	5	1	9	2	3
2	4	9	8	6	3	1	7	5
5	1	3	2	9	7	8	4	6
3	5	8	9	7	6	4	1	2
7	2	4	1	8	5	3	6	9
1	9	6	3	2	4	5	8	7

Sudoku # 263

1	5	8	9	4	3	6	7	2
3	6	4	1	2	7	9	8	5
7	9	2	5	6	8	3	1	4
5	8	9	4	7	2	1	6	3
2	3	7	6	8	1	5	4	9
4	1	6	3	9	5	7	2	8
6	7	3	2	5	4	8	9	1
8	2	5	7	1	9	4	3	6
9	4	1	8	3	6	2	5	7

Sudoku # 264

1	2	7	8	9	6	5	3	4
3	4	5	2	7	1	8	6	9
9	6	8	3	4	5	1	7	2
8	1	3	9	2	4	6	5	7
2	7	6	5	1	3	4	9	8
4	5	9	6	8	7	3	2	1
6	8	2	1	5	9	7	4	3
5	9	4	7	3	8	2	1	6
7	3	1	4	6	2	9	8	5

Sudoku # 265

2	5	9	7	8	6	1	3	4
7	1	6	2	4	3	9	5	8
8	4	3	5	1	9	7	2	6
6	3	4	8	5	7	2	1	9
5	8	1	3	9	2	6	4	7
9	2	7	4	6	1	5	8	3
3	7	8	9	2	5	4	6	1
1	9	2	6	3	4	8	7	5
4	6	5	1	7	8	3	9	2

Sudoku # 266

7	4	9	6	1	8	3	5	2
1	8	3	5	7	2	4	6	9
5	2	6	9	3	4	1	8	7
6	3	8	4	9	7	5	2	1
4	9	7	2	5	1	6	3	8
2	1	5	3	8	6	7	9	4
3	6	1	7	2	9	8	4	5
9	7	4	8	6	5	2	1	3
8	5	2	1	4	3	9	7	6

Sudoku # 267

5	8	3	2	6	9	7	4	1
7	2	1	5	3	4	9	8	6
9	4	6	7	1	8	3	5	2
6	3	9	4	5	7	1	2	8
4	1	8	6	9	2	5	7	3
2	7	5	1	8	3	6	9	4
8	9	2	3	7	1	4	6	5
3	5	7	8	4	6	2	1	9
1	6	4	9	2	5	8	3	7

Sudoku # 268

3	5	8	1	7	2	4	9	6
9	6	7	4	8	5	3	2	1
2	4	1	3	6	9	8	5	7
8	7	6	2	9	1	5	4	3
1	9	3	7	5	4	2	6	8
5	2	4	6	3	8	7	1	9
4	1	9	8	2	3	6	7	5
6	8	2	5	1	7	9	3	4
7	3	5	9	4	6	1	8	2

Sudoku # 269

4	9	2	5	7	6	8	1	3
1	6	7	4	8	3	2	9	5
8	5	3	9	1	2	6	4	7
9	2	8	3	6	7	1	5	4
6	4	5	2	9	1	3	7	8
3	7	1	8	4	5	9	2	6
2	1	6	7	3	4	5	8	9
7	3	9	1	5	8	4	6	2
5	8	4	6	2	9	7	3	1

Sudoku # 270

1	3	5	6	7	4	8	2	9
4	7	2	1	9	8	5	6	3
8	6	9	5	3	2	7	4	1
3	2	4	8	5	7	1	9	6
5	8	6	3	1	9	2	7	4
7	9	1	2	4	6	3	8	5
9	1	7	4	2	5	6	3	8
2	5	8	9	6	3	4	1	7
6	4	3	7	8	1	9	5	2

Sudoku # 271

6	8	5	3	9	7	1	2	4
1	7	9	5	4	2	3	6	8
2	3	4	6	8	1	7	5	9
8	2	6	7	1	5	4	9	3
5	4	3	8	2	9	6	7	1
7	9	1	4	6	3	5	8	2
3	5	8	2	7	4	9	1	6
9	6	7	1	3	8	2	4	5
4	1	2	9	5	6	8	3	7

Sudoku # 272

1	8	7	2	6	5	3	4	9
2	9	3	4	8	7	5	1	6
4	5	6	9	3	1	7	2	8
9	2	5	1	4	3	8	6	7
3	4	8	7	9	6	2	5	1
7	6	1	5	2	8	4	9	3
6	1	2	8	7	4	9	3	5
8	3	9	6	5	2	1	7	4
5	7	4	3	1	9	6	8	2

Sudoku # 273

8	6	5	9	1	3	2	7	4
9	4	7	2	6	5	8	1	3
3	1	2	8	4	7	5	9	6
1	9	3	5	8	6	7	4	2
6	5	4	7	3	2	9	8	1
2	7	8	1	9	4	3	6	5
4	2	9	3	7	1	6	5	8
5	8	1	6	2	9	4	3	7
7	3	6	4	5	8	1	2	9

Sudoku # 274

9	3	6	1	4	8	5	7	2
7	1	8	6	2	5	4	9	3
5	2	4	9	7	3	6	1	8
4	6	2	3	9	1	8	5	7
3	7	9	5	8	4	2	6	1
1	8	5	2	6	7	3	4	9
8	9	1	4	5	2	7	3	6
6	5	7	8	3	9	1	2	4
2	4	3	7	1	6	9	8	5

Sudoku # 275

1	7	9	6	3	4	2	5	8
2	5	4	9	1	8	7	6	3
6	8	3	2	7	5	9	4	1
5	9	2	4	6	1	3	8	7
4	3	1	7	8	2	6	9	5
8	6	7	3	5	9	1	2	4
3	2	8	5	9	7	4	1	6
9	1	6	8	4	3	5	7	2
7	4	5	1	2	6	8	3	9

Sudoku # 276

2	5	1	8	6	7	9	3	4
6	8	7	3	4	9	2	5	1
3	9	4	5	1	2	7	8	6
4	6	3	2	9	1	8	7	5
9	7	2	6	8	5	1	4	3
8	1	5	4	7	3	6	9	2
1	2	9	7	3	4	5	6	8
5	4	6	9	2	8	3	1	7
7	3	8	1	5	6	4	2	9

Sudoku # 277

3	6	8	4	2	5	9	7	1
2	7	9	8	6	1	5	4	3
5	4	1	7	9	3	2	6	8
8	1	7	9	5	4	3	2	6
6	3	2	1	8	7	4	9	5
4	9	5	6	3	2	1	8	7
7	2	6	5	1	9	8	3	4
1	8	3	2	4	6	7	5	9
9	5	4	3	7	8	6	1	2

Sudoku # 278

8	6	3	5	7	4	2	1	9
5	9	7	2	6	1	8	4	3
4	2	1	3	9	8	6	5	7
7	8	9	4	1	2	3	6	5
3	1	2	9	5	6	4	7	8
6	4	5	7	8	3	9	2	1
9	5	8	6	2	7	1	3	4
1	3	6	8	4	5	7	9	2
2	7	4	1	3	9	5	8	6

Sudoku # 279

9	6	8	2	3	5	4	7	1
3	2	1	8	4	7	5	9	6
7	5	4	6	1	9	8	2	3
4	1	3	5	6	2	9	8	7
5	9	2	3	7	8	1	6	4
6	8	7	4	9	1	3	5	2
1	4	5	7	8	6	2	3	9
8	7	9	1	2	3	6	4	5
2	3	6	9	5	4	7	1	8

Sudoku # 280

9	3	7	5	2	4	1	6	8
5	8	4	3	1	6	9	2	7
1	2	6	7	9	8	3	4	5
6	1	9	8	5	2	4	7	3
2	7	8	9	4	3	6	5	1
3	4	5	6	7	1	8	9	2
4	6	2	1	8	7	5	3	9
7	5	1	4	3	9	2	8	6
8	9	3	2	6	5	7	1	4

Sudoku # 281

1	2	7	3	4	8	9	6	5
4	9	3	6	1	5	7	2	8
6	8	5	7	2	9	4	1	3
9	7	4	5	8	1	2	3	6
5	6	1	4	3	2	8	9	7
8	3	2	9	7	6	5	4	1
7	5	6	2	9	3	1	8	4
2	4	8	1	6	7	3	5	9
3	1	9	8	5	4	6	7	2

Sudoku # 282

6	9	5	7	4	8	2	1	3
7	1	8	3	5	2	4	6	9
4	3	2	9	6	1	8	5	7
5	4	3	2	8	7	1	9	6
2	6	9	1	3	4	7	8	5
1	8	7	6	9	5	3	4	2
3	5	1	8	7	9	6	2	4
9	2	6	4	1	3	5	7	8
8	7	4	5	2	6	9	3	1

Sudoku # 283

4	2	7	6	8	3	5	9	1
9	1	8	4	5	2	6	3	7
5	6	3	7	1	9	8	2	4
7	3	6	2	9	5	4	1	8
8	9	1	3	6	4	2	7	5
2	4	5	1	7	8	9	6	3
3	7	2	5	4	6	1	8	9
1	8	4	9	2	7	3	5	6
6	5	9	8	3	1	7	4	2

Sudoku # 284

4	2	5	8	1	3	7	9	6
3	6	1	2	9	7	8	5	4
8	9	7	4	6	5	1	2	3
7	8	4	5	3	6	2	1	9
1	3	9	7	8	2	4	6	5
6	5	2	1	4	9	3	7	8
2	7	8	9	5	4	6	3	1
5	4	6	3	7	1	9	8	2
9	1	3	6	2	8	5	4	7

Sudoku # 285

8	2	4	3	6	9	1	7	5
7	6	3	5	1	4	8	9	2
5	9	1	2	8	7	3	4	6
1	3	2	4	7	8	6	5	9
4	8	9	6	3	5	7	2	1
6	7	5	9	2	1	4	8	3
2	1	8	7	5	3	9	6	4
3	4	6	8	9	2	5	1	7
9	5	7	1	4	6	2	3	8

Sudoku # 286

2	7	4	9	3	1	6	8	5
8	3	9	5	4	6	2	1	7
5	6	1	7	2	8	9	3	4
3	8	7	2	5	9	1	4	6
9	2	6	3	1	4	5	7	8
4	1	5	6	8	7	3	2	9
1	9	8	4	6	2	7	5	3
6	5	2	8	7	3	4	9	1
7	4	3	1	9	5	8	6	2

Sudoku # 287

5	3	1	9	4	2	8	7	6
4	6	8	1	3	7	2	9	5
7	2	9	8	6	5	1	3	4
8	9	3	6	2	4	7	5	1
1	4	5	3	7	9	6	8	2
2	7	6	5	8	1	3	4	9
3	5	2	7	9	6	4	1	8
9	8	4	2	1	3	5	6	7
6	1	7	4	5	8	9	2	3

Sudoku # 288

5	4	2	7	8	3	9	6	1
3	9	8	1	5	6	2	7	4
7	6	1	4	9	2	3	5	8
1	3	6	9	2	8	7	4	5
9	8	5	6	4	7	1	2	3
4	2	7	5	3	1	8	9	6
8	7	9	3	6	5	4	1	2
2	5	4	8	1	9	6	3	7
6	1	3	2	7	4	5	8	9

Sudoku # 289

3	7	2	1	9	6	8	4	5
9	5	1	2	4	8	3	6	7
4	6	8	5	3	7	1	9	2
5	2	6	4	8	1	7	3	9
8	1	3	6	7	9	2	5	4
7	4	9	3	2	5	6	1	8
1	8	5	9	6	2	4	7	3
2	9	4	7	1	3	5	8	6
6	3	7	8	5	4	9	2	1

Sudoku # 290

7	2	9	3	4	6	5	1	8
3	5	6	7	8	1	2	4	9
1	8	4	5	2	9	3	6	7
5	3	1	8	7	2	4	9	6
2	9	8	1	6	4	7	3	5
6	4	7	9	5	3	8	2	1
8	1	3	4	9	5	6	7	2
4	7	2	6	1	8	9	5	3
9	6	5	2	3	7	1	8	4

Sudoku # 291

5	7	4	3	9	2	1	6	8
1	2	6	7	4	8	9	5	3
9	8	3	1	6	5	7	4	2
4	1	8	6	7	3	5	2	9
7	5	2	9	8	4	3	1	6
6	3	9	5	2	1	4	8	7
2	9	1	4	3	6	8	7	5
3	6	5	8	1	7	2	9	4
8	4	7	2	5	9	6	3	1

Sudoku # 292

2	5	9	7	8	3	6	4	1
4	3	7	1	6	5	2	9	8
1	8	6	9	4	2	5	7	3
9	1	5	8	2	7	3	6	4
3	2	4	6	9	1	7	8	5
7	6	8	5	3	4	9	1	2
5	9	3	4	1	6	8	2	7
8	4	2	3	7	9	1	5	6
6	7	1	2	5	8	4	3	9

Sudoku # 293

5	7	1	2	6	9	8	4	3
2	8	9	1	3	4	6	5	7
4	3	6	7	5	8	2	1	9
3	4	5	8	7	2	1	9	6
1	9	8	3	4	6	7	2	5
6	2	7	9	1	5	3	8	4
7	1	2	4	9	3	5	6	8
8	5	4	6	2	7	9	3	1
9	6	3	5	8	1	4	7	2

Sudoku # 294

7	3	8	9	4	1	6	5	2
6	5	2	7	3	8	9	4	1
9	1	4	5	6	2	7	3	8
8	4	3	1	2	9	5	6	7
2	7	1	3	5	6	4	8	9
5	9	6	4	8	7	2	1	3
3	8	7	6	9	5	1	2	4
4	6	9	2	1	3	8	7	5
1	2	5	8	7	4	3	9	6

Sudoku # 295

2	4	5	9	1	7	3	8	6
6	1	7	4	8	3	5	2	9
9	3	8	6	5	2	1	7	4
1	8	2	7	3	4	9	6	5
3	7	6	8	9	5	2	4	1
4	5	9	1	2	6	7	3	8
8	9	3	2	4	1	6	5	7
7	2	1	5	6	8	4	9	3
5	6	4	3	7	9	8	1	2

Sudoku # 296

3	4	1	9	2	5	8	6	7
6	5	8	7	3	1	2	4	9
7	2	9	4	6	8	1	5	3
9	7	2	1	5	4	3	8	6
4	6	3	2	8	7	9	1	5
8	1	5	3	9	6	4	7	2
2	8	7	6	1	9	5	3	4
5	3	6	8	4	2	7	9	1
1	9	4	5	7	3	6	2	8

Sudoku # 297

4	7	8	5	1	3	2	6	9
5	3	6	2	8	9	4	7	1
1	2	9	4	7	6	8	3	5
9	8	2	7	6	4	5	1	3
7	5	4	8	3	1	9	2	6
6	1	3	9	2	5	7	8	4
3	4	7	1	5	8	6	9	2
2	9	1	6	4	7	3	5	8
8	6	5	3	9	2	1	4	7

Sudoku # 298

6	2	7	5	1	4	9	8	3
9	5	3	6	8	2	1	4	7
8	1	4	7	3	9	5	6	2
4	3	9	2	6	8	7	5	1
7	6	5	4	9	1	3	2	8
1	8	2	3	5	7	4	9	6
3	7	6	8	4	5	2	1	9
2	4	1	9	7	6	8	3	5
5	9	8	1	2	3	6	7	4

Sudoku # 299

8	9	7	1	2	4	5	6	3
5	4	2	9	6	3	7	1	8
6	3	1	8	5	7	2	9	4
1	5	8	3	7	2	9	4	6
2	6	9	5	4	1	8	3	7
4	7	3	6	8	9	1	5	2
3	1	6	7	9	8	4	2	5
7	2	5	4	1	6	3	8	9
9	8	4	2	3	5	6	7	1

Sudoku # 300

6	8	1	4	2	5	7	9	3
2	7	4	6	3	9	1	8	5
9	5	3	7	8	1	6	4	2
3	6	7	2	1	8	9	5	4
4	9	5	3	6	7	2	1	8
1	2	8	9	5	4	3	6	7
8	1	6	5	7	2	4	3	9
5	4	2	1	9	3	8	7	6
7	3	9	8	4	6	5	2	1

Sudoku # 301

7	1	5	6	4	8	9	3	2
3	2	9	1	7	5	6	8	4
4	6	8	2	3	9	7	1	5
9	7	4	8	5	1	3	2	6
8	3	1	7	2	6	5	4	9
6	5	2	3	9	4	8	7	1
5	8	7	4	6	2	1	9	3
2	9	3	5	1	7	4	6	8
1	4	6	9	8	3	2	5	7

Sudoku # 302

1	3	9	8	5	2	6	4	7
6	2	8	9	7	4	5	3	1
5	7	4	1	3	6	8	9	2
7	9	3	4	2	8	1	6	5
2	8	1	3	6	5	4	7	9
4	5	6	7	9	1	2	8	3
8	1	7	2	4	9	3	5	6
3	4	5	6	1	7	9	2	8
9	6	2	5	8	3	7	1	4

Sudoku # 303

1	9	6	7	2	8	4	3	5
2	8	5	3	6	4	9	7	1
7	3	4	1	5	9	6	2	8
8	2	3	5	1	6	7	9	4
6	5	7	9	4	2	8	1	3
9	4	1	8	7	3	2	5	6
3	6	2	4	9	1	5	8	7
4	7	8	2	3	5	1	6	9
5	1	9	6	8	7	3	4	2

Sudoku # 304

6	2	7	9	3	4	5	8	1
9	3	1	6	8	5	4	7	2
5	8	4	2	1	7	6	9	3
4	9	8	7	6	1	2	3	5
1	6	2	8	5	3	7	4	9
3	7	5	4	2	9	8	1	6
8	1	9	5	7	2	3	6	4
7	5	3	1	4	6	9	2	8
2	4	6	3	9	8	1	5	7

Sudoku # 305

5	9	4	7	3	6	8	1	2
3	1	6	9	2	8	4	5	7
7	8	2	5	1	4	3	6	9
8	6	3	1	7	5	2	9	4
9	7	1	3	4	2	6	8	5
2	4	5	8	6	9	7	3	1
4	2	8	6	9	1	5	7	3
6	3	9	2	5	7	1	4	8
1	5	7	4	8	3	9	2	6

Sudoku # 306

9	2	7	4	5	1	3	8	6
6	4	3	7	8	9	1	5	2
8	1	5	2	3	6	7	4	9
1	6	8	5	7	3	9	2	4
4	7	9	6	1	2	8	3	5
5	3	2	9	4	8	6	7	1
3	9	4	8	6	5	2	1	7
7	8	6	1	2	4	5	9	3
2	5	1	3	9	7	4	6	8

Sudoku # 307

6	4	8	9	1	3	2	7	5
1	5	2	4	6	7	3	8	9
7	9	3	8	5	2	4	6	1
3	7	5	2	9	4	6	1	8
9	8	1	6	3	5	7	2	4
4	2	6	7	8	1	5	9	3
8	6	4	3	7	9	1	5	2
5	3	7	1	2	8	9	4	6
2	1	9	5	4	6	8	3	7

Sudoku # 308

3	5	6	1	8	7	9	4	2
2	7	8	4	3	9	6	1	5
9	1	4	5	6	2	3	7	8
4	9	7	3	5	1	8	2	6
6	8	2	9	7	4	5	3	1
1	3	5	8	2	6	4	9	7
5	4	3	7	1	8	2	6	9
7	6	9	2	4	5	1	8	3
8	2	1	6	9	3	7	5	4

Sudoku # 309

6	8	5	9	3	1	7	4	2
1	4	9	2	6	7	3	8	5
3	7	2	8	5	4	9	1	6
7	1	8	6	9	3	5	2	4
9	5	3	1	4	2	8	6	7
4	2	6	7	8	5	1	9	3
2	9	1	3	7	6	4	5	8
8	3	4	5	2	9	6	7	1
5	6	7	4	1	8	2	3	9

Sudoku # 310

6	3	8	9	2	7	5	4	1
4	2	7	3	1	5	9	8	6
9	5	1	8	4	6	7	2	3
8	7	6	1	5	9	2	3	4
3	9	2	4	6	8	1	7	5
5	1	4	7	3	2	6	9	8
1	8	9	6	7	3	4	5	2
2	4	3	5	9	1	8	6	7
7	6	5	2	8	4	3	1	9

Sudoku # 311

7	8	6	4	9	1	5	2	3
2	4	5	3	8	7	9	6	1
9	1	3	5	6	2	4	7	8
6	5	1	2	4	9	8	3	7
4	9	7	8	1	3	6	5	2
3	2	8	6	7	5	1	4	9
8	3	2	1	5	4	7	9	6
5	6	9	7	3	8	2	1	4
1	7	4	9	2	6	3	8	5

Sudoku # 312

9	3	1	4	8	5	2	6	7
7	6	4	1	3	2	9	5	8
5	8	2	7	9	6	3	4	1
4	7	8	6	5	9	1	2	3
6	9	5	2	1	3	7	8	4
2	1	3	8	7	4	6	9	5
1	4	9	3	2	8	5	7	6
3	2	6	5	4	7	8	1	9
8	5	7	9	6	1	4	3	2

Sudoku # 313

4	9	7	5	8	3	6	2	1
3	5	2	9	6	1	8	7	4
8	6	1	7	4	2	5	9	3
7	3	8	4	5	9	1	6	2
1	4	5	3	2	6	9	8	7
6	2	9	8	1	7	3	4	5
9	8	3	1	7	4	2	5	6
2	1	4	6	9	5	7	3	8
5	7	6	2	3	8	4	1	9

Sudoku # 314

7	1	2	4	9	3	8	5	6
6	8	4	1	5	7	3	2	9
5	9	3	6	2	8	7	1	4
3	4	6	7	1	5	9	8	2
2	7	9	3	8	6	1	4	5
1	5	8	2	4	9	6	3	7
8	6	7	5	3	2	4	9	1
4	3	5	9	6	1	2	7	8
9	2	1	8	7	4	5	6	3

Sudoku # 315

9	3	8	1	5	6	4	7	2
7	4	6	8	2	9	3	1	5
2	1	5	4	7	3	9	8	6
5	7	1	3	8	4	2	6	9
3	9	4	6	1	2	8	5	7
6	8	2	5	9	7	1	4	3
8	2	9	7	6	1	5	3	4
1	6	3	9	4	5	7	2	8
4	5	7	2	3	8	6	9	1

Sudoku # 316

2	6	9	3	4	1	8	7	5
3	8	5	7	2	9	4	6	1
1	4	7	5	8	6	2	9	3
9	5	1	2	6	4	7	3	8
8	7	4	1	9	3	5	2	6
6	3	2	8	7	5	9	1	4
4	2	3	9	1	8	6	5	7
7	1	6	4	5	2	3	8	9
5	9	8	6	3	7	1	4	2

Sudoku # 317

9	7	2	1	6	8	3	4	5
6	1	5	7	3	4	9	8	2
8	4	3	2	9	5	1	7	6
7	8	4	5	1	3	6	2	9
5	9	6	8	4	2	7	3	1
3	2	1	6	7	9	8	5	4
2	6	9	3	5	7	4	1	8
4	5	7	9	8	1	2	6	3
1	3	8	4	2	6	5	9	7

Sudoku # 318

6	7	2	8	1	4	9	3	5
8	5	4	9	6	3	1	7	2
1	9	3	7	2	5	8	4	6
5	8	9	6	4	1	3	2	7
2	3	7	5	9	8	4	6	1
4	1	6	2	3	7	5	8	9
3	6	5	4	7	9	2	1	8
9	2	1	3	8	6	7	5	4
7	4	8	1	5	2	6	9	3

Sudoku # 319

6	2	5	4	9	3	7	1	8
8	4	1	7	2	6	3	5	9
3	7	9	8	1	5	4	6	2
7	3	4	1	5	2	9	8	6
9	6	2	3	8	4	1	7	5
5	1	8	9	6	7	2	3	4
1	5	6	2	3	9	8	4	7
2	8	7	5	4	1	6	9	3
4	9	3	6	7	8	5	2	1

Sudoku # 320

3	8	4	1	2	5	9	6	7
1	9	6	7	4	3	8	5	2
2	5	7	6	8	9	4	3	1
5	6	3	4	1	2	7	8	9
7	1	9	3	5	8	2	4	6
8	4	2	9	7	6	5	1	3
9	7	8	5	6	1	3	2	4
4	2	1	8	3	7	6	9	5
6	3	5	2	9	4	1	7	8

Sudoku # 321

6	1	4	7	8	3	9	5	2
7	9	8	4	5	2	3	6	1
2	3	5	6	1	9	8	7	4
5	2	3	9	4	8	6	1	7
8	7	9	5	6	1	4	2	3
4	6	1	2	3	7	5	8	9
9	4	6	1	2	5	7	3	8
3	5	2	8	7	4	1	9	6
1	8	7	3	9	6	2	4	5

Sudoku # 322

2	5	1	3	8	7	6	9	4
9	4	3	1	6	2	7	5	8
6	8	7	5	9	4	2	3	1
1	7	8	9	4	6	5	2	3
3	6	2	8	1	5	9	4	7
4	9	5	2	7	3	8	1	6
7	2	9	6	3	1	4	8	5
5	1	6	4	2	8	3	7	9
8	3	4	7	5	9	1	6	2

Sudoku # 323

9	6	8	3	4	5	1	7	2
1	7	2	6	9	8	4	5	3
5	3	4	7	1	2	8	9	6
7	1	5	8	2	9	3	6	4
4	8	9	1	6	3	5	2	7
3	2	6	4	5	7	9	1	8
2	9	7	5	3	4	6	8	1
8	4	1	9	7	6	2	3	5
6	5	3	2	8	1	7	4	9

Sudoku # 324

3	4	5	6	8	7	9	1	2
8	6	9	4	1	2	3	7	5
7	2	1	9	3	5	8	4	6
1	9	3	7	6	4	2	5	8
5	7	6	3	2	8	4	9	1
4	8	2	1	5	9	7	6	3
9	1	8	2	7	6	5	3	4
2	3	7	5	4	1	6	8	9
6	5	4	8	9	3	1	2	7

Sudoku # 325

7	8	2	4	9	6	5	1	3
4	6	9	5	1	3	7	2	8
5	1	3	8	7	2	9	6	4
9	4	6	1	3	7	8	5	2
3	5	8	9	2	4	6	7	1
2	7	1	6	8	5	3	4	9
1	2	7	3	6	9	4	8	5
8	9	4	7	5	1	2	3	6
6	3	5	2	4	8	1	9	7

Sudoku # 326

1	4	6	2	7	3	8	9	5
3	8	7	9	5	6	4	2	1
9	2	5	8	4	1	7	3	6
6	1	3	5	9	8	2	7	4
2	9	4	1	6	7	3	5	8
7	5	8	3	2	4	6	1	9
4	6	2	7	1	9	5	8	3
8	7	1	4	3	5	9	6	2
5	3	9	6	8	2	1	4	7

Sudoku # 327

3	1	7	9	2	5	8	4	6
8	6	9	4	3	1	5	7	2
2	4	5	7	8	6	1	9	3
1	2	4	5	7	9	3	6	8
7	9	8	3	6	2	4	1	5
5	3	6	1	4	8	7	2	9
4	7	2	6	5	3	9	8	1
9	8	3	2	1	4	6	5	7
6	5	1	8	9	7	2	3	4

Sudoku # 328

5	7	3	8	4	9	2	6	1
6	2	9	3	5	1	7	8	4
1	8	4	6	7	2	3	5	9
3	5	7	2	9	6	4	1	8
2	1	8	5	3	4	6	9	7
4	9	6	1	8	7	5	3	2
8	6	1	7	2	3	9	4	5
7	4	5	9	6	8	1	2	3
9	3	2	4	1	5	8	7	6

Sudoku # 329

6	9	7	3	5	8	4	2	1
5	2	4	7	9	1	8	3	6
1	3	8	6	2	4	5	9	7
4	7	3	8	6	2	9	1	5
8	1	6	9	4	5	3	7	2
9	5	2	1	3	7	6	8	4
2	8	5	4	7	9	1	6	3
3	4	9	2	1	6	7	5	8
7	6	1	5	8	3	2	4	9

Sudoku # 330

9	1	2	3	8	7	6	4	5
3	4	6	9	5	2	7	1	8
8	7	5	1	6	4	3	9	2
7	9	8	4	1	6	5	2	3
5	6	1	7	2	3	4	8	9
2	3	4	8	9	5	1	6	7
1	8	3	6	7	9	2	5	4
6	5	7	2	4	8	9	3	1
4	2	9	5	3	1	8	7	6

Sudoku # 331

3	6	5	4	2	7	8	9	1
2	9	8	5	1	6	4	3	7
7	1	4	8	9	3	2	5	6
6	7	9	3	5	8	1	4	2
5	2	3	6	4	1	7	8	9
8	4	1	9	7	2	3	6	5
4	8	2	1	6	5	9	7	3
9	5	7	2	3	4	6	1	8
1	3	6	7	8	9	5	2	4

Sudoku # 332

3	9	8	5	7	4	6	2	1
1	4	6	8	3	2	7	5	9
2	5	7	6	9	1	8	3	4
4	6	1	9	5	8	3	7	2
9	8	3	2	4	7	1	6	5
5	7	2	3	1	6	9	4	8
6	1	5	7	2	9	4	8	3
8	3	9	4	6	5	2	1	7
7	2	4	1	8	3	5	9	6

Sudoku # 333

1	5	6	2	8	9	3	4	7
4	8	3	5	6	7	1	9	2
9	2	7	1	4	3	5	8	6
3	4	5	7	1	6	8	2	9
7	9	1	4	2	8	6	5	3
2	6	8	3	9	5	7	1	4
6	7	4	9	5	1	2	3	8
8	1	9	6	3	2	4	7	5
5	3	2	8	7	4	9	6	1

Sudoku # 334

3	5	8	4	7	9	2	1	6
4	2	9	3	1	6	5	8	7
6	7	1	8	2	5	4	3	9
9	8	3	5	4	2	7	6	1
5	1	4	9	6	7	3	2	8
7	6	2	1	3	8	9	4	5
2	3	7	6	9	1	8	5	4
1	4	5	7	8	3	6	9	2
8	9	6	2	5	4	1	7	3

Sudoku # 335

4	2	1	8	3	5	6	7	9
5	7	9	4	6	1	3	2	8
6	3	8	9	7	2	4	5	1
3	5	2	1	4	9	8	6	7
1	4	7	5	8	6	2	9	3
8	9	6	3	2	7	5	1	4
9	8	5	6	1	4	7	3	2
2	6	3	7	9	8	1	4	5
7	1	4	2	5	3	9	8	6

Sudoku # 336

1	5	8	7	2	6	9	4	3
6	3	4	9	5	8	2	7	1
2	7	9	4	1	3	8	5	6
7	1	3	6	8	9	5	2	4
8	2	5	3	4	1	7	6	9
4	9	6	5	7	2	3	1	8
5	8	1	2	9	4	6	3	7
3	4	7	8	6	5	1	9	2
9	6	2	1	3	7	4	8	5

Sudoku # 337

7	2	3	9	1	6	4	5	8
8	1	5	4	2	3	6	7	9
6	9	4	5	8	7	2	3	1
4	5	2	7	6	1	8	9	3
1	6	7	3	9	8	5	4	2
3	8	9	2	5	4	1	6	7
5	3	8	6	7	2	9	1	4
9	4	1	8	3	5	7	2	6
2	7	6	1	4	9	3	8	5

Sudoku # 338

4	1	7	3	5	8	2	9	6
8	9	3	1	6	2	4	5	7
2	5	6	9	4	7	3	8	1
1	6	2	8	3	9	5	7	4
5	7	4	2	1	6	8	3	9
9	3	8	5	7	4	6	1	2
7	2	1	4	8	5	9	6	3
3	8	9	6	2	1	7	4	5
6	4	5	7	9	3	1	2	8

Sudoku # 339

7	4	5	8	2	9	3	1	6
1	2	9	3	5	6	8	7	4
6	3	8	1	4	7	5	9	2
5	7	1	6	3	4	9	2	8
3	9	6	7	8	2	4	5	1
4	8	2	9	1	5	7	6	3
2	1	4	5	9	3	6	8	7
8	5	7	4	6	1	2	3	9
9	6	3	2	7	8	1	4	5

Sudoku # 340

5	1	9	4	8	2	7	6	3
3	2	8	1	7	6	4	9	5
6	4	7	9	3	5	2	8	1
2	8	3	5	1	9	6	7	4
7	6	4	3	2	8	1	5	9
1	9	5	6	4	7	3	2	8
4	7	2	8	5	3	9	1	6
8	3	6	2	9	1	5	4	7
9	5	1	7	6	4	8	3	2

Sudoku # 341

5	1	8	2	4	3	7	9	6
4	6	7	8	5	9	2	3	1
9	2	3	7	1	6	5	8	4
2	9	4	3	7	5	1	6	8
8	5	6	1	9	4	3	7	2
7	3	1	6	2	8	4	5	9
6	4	2	5	8	7	9	1	3
3	7	9	4	6	1	8	2	5
1	8	5	9	3	2	6	4	7

Sudoku # 342

2	4	7	6	3	1	8	5	9
5	8	9	2	7	4	1	6	3
6	1	3	5	9	8	7	4	2
9	3	5	4	2	7	6	8	1
8	7	4	9	1	6	2	3	5
1	2	6	3	8	5	4	9	7
7	6	2	8	5	9	3	1	4
4	5	1	7	6	3	9	2	8
3	9	8	1	4	2	5	7	6

Sudoku # 343

8	1	6	2	4	5	7	3	9
5	9	2	6	7	3	8	4	1
4	7	3	8	9	1	5	6	2
7	2	1	9	6	4	3	5	8
9	3	5	1	8	7	4	2	6
6	8	4	5	3	2	9	1	7
1	4	9	7	5	6	2	8	3
3	6	7	4	2	8	1	9	5
2	5	8	3	1	9	6	7	4

Sudoku # 344

9	5	6	7	2	3	8	1	4
3	2	8	6	1	4	7	5	9
1	7	4	8	9	5	3	6	2
5	4	7	2	6	1	9	3	8
8	1	9	5	3	7	2	4	6
2	6	3	9	4	8	1	7	5
7	3	2	4	8	6	5	9	1
4	8	5	1	7	9	6	2	3
6	9	1	3	5	2	4	8	7

Sudoku # 345

1	6	3	9	7	2	8	4	5
9	8	5	3	6	4	2	7	1
4	7	2	5	8	1	3	6	9
6	5	4	8	1	3	7	9	2
3	1	8	7	2	9	4	5	6
7	2	9	4	5	6	1	8	3
2	9	7	1	4	5	6	3	8
8	3	1	6	9	7	5	2	4
5	4	6	2	3	8	9	1	7

Sudoku # 346

2	6	4	9	1	7	5	8	3
8	1	5	3	2	6	9	7	4
9	7	3	8	4	5	6	2	1
5	9	1	4	7	2	8	3	6
3	8	6	5	9	1	2	4	7
7	4	2	6	3	8	1	5	9
6	5	7	1	8	3	4	9	2
1	3	9	2	5	4	7	6	8
4	2	8	7	6	9	3	1	5

Sudoku # 347

8	7	3	5	4	9	2	1	6
4	9	5	1	6	2	7	3	8
1	2	6	3	8	7	5	4	9
7	4	9	8	5	6	1	2	3
5	3	1	2	9	4	6	8	7
2	6	8	7	3	1	9	5	4
6	8	7	4	2	5	3	9	1
3	1	2	9	7	8	4	6	5
9	5	4	6	1	3	8	7	2

Sudoku # 348

9	4	5	3	8	6	1	2	7
7	2	8	5	9	1	3	6	4
1	3	6	4	7	2	8	5	9
5	9	7	1	2	8	4	3	6
8	6	4	7	5	3	9	1	2
2	1	3	6	4	9	5	7	8
6	8	1	2	3	4	7	9	5
3	7	9	8	6	5	2	4	1
4	5	2	9	1	7	6	8	3

Sudoku # 349

3	4	6	9	1	7	5	8	2
2	7	1	3	5	8	4	9	6
8	5	9	6	4	2	7	3	1
7	6	4	2	9	3	8	1	5
9	8	3	5	7	1	6	2	4
1	2	5	8	6	4	9	7	3
4	3	7	1	8	6	2	5	9
5	1	8	4	2	9	3	6	7
6	9	2	7	3	5	1	4	8

Sudoku # 350

6	7	5	1	2	4	3	8	9
1	9	4	5	8	3	6	7	2
3	2	8	9	7	6	1	5	4
7	6	9	2	1	8	4	3	5
4	8	3	6	5	9	2	1	7
2	5	1	4	3	7	8	9	6
5	3	2	7	6	1	9	4	8
8	4	7	3	9	2	5	6	1
9	1	6	8	4	5	7	2	3

Sudoku # 351

1	7	9	5	6	8	3	2	4
5	6	8	2	3	4	7	9	1
2	3	4	1	9	7	5	8	6
6	5	2	8	7	1	4	3	9
9	1	7	3	4	5	8	6	2
8	4	3	9	2	6	1	7	5
4	8	6	7	1	2	9	5	3
3	2	5	4	8	9	6	1	7
7	9	1	6	5	3	2	4	8

Sudoku # 352

4	6	8	9	1	5	2	3	7
9	3	1	6	2	7	4	8	5
7	2	5	4	3	8	9	1	6
2	4	3	8	7	1	6	5	9
6	8	9	3	5	4	7	2	1
5	1	7	2	6	9	8	4	3
1	5	6	7	4	2	3	9	8
8	7	2	1	9	3	5	6	4
3	9	4	5	8	6	1	7	2

Sudoku # 353

5	3	6	7	9	8	4	2	1
9	7	2	6	1	4	3	5	8
8	4	1	5	3	2	7	6	9
7	2	4	9	6	5	8	1	3
3	9	8	1	2	7	6	4	5
6	1	5	8	4	3	2	9	7
4	5	7	2	8	1	9	3	6
1	6	3	4	7	9	5	8	2
2	8	9	3	5	6	1	7	4

Sudoku # 354

4	1	2	7	9	5	6	8	3
3	7	6	4	2	8	9	5	1
5	9	8	1	6	3	7	2	4
2	6	1	9	3	7	5	4	8
7	4	5	6	8	2	3	1	9
8	3	9	5	1	4	2	7	6
6	8	7	2	4	9	1	3	5
1	2	3	8	5	6	4	9	7
9	5	4	3	7	1	8	6	2

Sudoku # 355

5	7	1	3	9	8	6	2	4
6	2	9	1	7	4	8	5	3
3	8	4	2	5	6	7	1	9
7	5	3	6	4	9	1	8	2
2	4	6	5	8	1	9	3	7
9	1	8	7	2	3	5	4	6
8	3	2	9	6	5	4	7	1
4	9	7	8	1	2	3	6	5
1	6	5	4	3	7	2	9	8

Sudoku # 356

7	9	3	4	8	5	2	1	6
2	8	1	6	9	7	4	3	5
6	4	5	3	2	1	8	9	7
5	1	6	9	3	8	7	2	4
3	2	8	1	7	4	5	6	9
4	7	9	5	6	2	3	8	1
9	5	2	7	1	3	6	4	8
1	3	7	8	4	6	9	5	2
8	6	4	2	5	9	1	7	3

Sudoku # 357

1	6	7	5	9	4	3	2	8
2	8	4	7	6	3	5	1	9
5	9	3	1	8	2	4	7	6
9	4	2	8	5	6	1	3	7
6	3	1	2	7	9	8	5	4
8	7	5	4	3	1	9	6	2
3	2	8	6	4	5	7	9	1
7	1	9	3	2	8	6	4	5
4	5	6	9	1	7	2	8	3

Sudoku # 358

5	4	1	9	3	6	2	7	8
3	7	8	4	2	5	6	9	1
9	6	2	1	7	8	4	3	5
1	5	9	6	4	7	3	8	2
6	3	7	8	5	2	1	4	9
2	8	4	3	1	9	7	5	6
4	2	3	5	9	1	8	6	7
8	1	5	7	6	3	9	2	4
7	9	6	2	8	4	5	1	3

Sudoku # 359

9	5	4	3	2	1	6	8	7
6	3	8	7	5	4	1	2	9
7	2	1	8	6	9	4	5	3
2	9	6	4	1	8	3	7	5
5	1	7	9	3	2	8	4	6
8	4	3	6	7	5	9	1	2
1	8	2	5	9	6	7	3	4
3	6	5	1	4	7	2	9	8
4	7	9	2	8	3	5	6	1

Sudoku # 360

3	1	9	7	4	6	2	8	5
7	4	6	8	5	2	9	3	1
2	8	5	1	9	3	6	4	7
8	2	1	9	6	4	5	7	3
4	6	7	5	3	1	8	2	9
9	5	3	2	7	8	4	1	6
6	9	8	3	2	7	1	5	4
1	7	4	6	8	5	3	9	2
5	3	2	4	1	9	7	6	8

Sudoku # 361

8	7	3	5	2	4	1	9	6
1	9	4	6	7	3	8	5	2
5	2	6	8	1	9	7	4	3
6	3	8	7	5	1	9	2	4
7	1	2	9	4	6	5	3	8
9	4	5	3	8	2	6	1	7
2	6	7	4	9	5	3	8	1
3	5	1	2	6	8	4	7	9
4	8	9	1	3	7	2	6	5

Sudoku # 362

5	2	3	6	9	8	7	4	1
7	4	8	1	3	2	5	6	9
1	9	6	7	5	4	8	3	2
9	8	2	5	6	1	3	7	4
6	7	5	3	4	9	2	1	8
3	1	4	2	8	7	6	9	5
8	3	9	4	2	6	1	5	7
2	6	1	9	7	5	4	8	3
4	5	7	8	1	3	9	2	6

Sudoku # 363

3	5	8	2	4	7	1	9	6
2	7	6	1	3	9	8	5	4
9	1	4	5	6	8	2	3	7
8	3	1	7	9	2	4	6	5
4	9	7	8	5	6	3	2	1
5	6	2	4	1	3	9	7	8
6	2	5	3	8	1	7	4	9
7	8	9	6	2	4	5	1	3
1	4	3	9	7	5	6	8	2

Sudoku # 364

3	5	2	9	6	8	4	7	1
4	7	8	1	5	2	3	9	6
1	6	9	7	3	4	2	8	5
6	8	5	4	2	9	1	3	7
9	1	7	5	8	3	6	2	4
2	4	3	6	7	1	9	5	8
7	9	1	3	4	5	8	6	2
8	3	6	2	1	7	5	4	9
5	2	4	8	9	6	7	1	3

Sudoku # 365

1	3	6	9	2	5	7	8	4
2	4	7	8	1	6	3	5	9
5	9	8	7	4	3	6	2	1
6	5	9	3	8	2	4	1	7
8	7	3	1	9	4	5	6	2
4	1	2	5	6	7	9	3	8
9	8	5	4	3	1	2	7	6
3	6	4	2	7	8	1	9	5
7	2	1	6	5	9	8	4	3

Sudoku # 366

8	7	5	4	6	1	2	9	3
1	4	6	2	9	3	8	7	5
2	3	9	5	7	8	4	6	1
4	1	3	9	8	7	6	5	2
6	2	8	3	5	4	9	1	7
9	5	7	1	2	6	3	8	4
3	9	2	8	1	5	7	4	6
5	6	4	7	3	9	1	2	8
7	8	1	6	4	2	5	3	9

Sudoku # 367

8	4	6	7	1	9	3	2	5
1	7	2	5	8	3	4	9	6
5	9	3	4	6	2	8	1	7
4	2	7	1	5	6	9	3	8
6	5	8	3	9	4	2	7	1
9	3	1	2	7	8	6	5	4
7	6	9	8	3	1	5	4	2
2	8	5	9	4	7	1	6	3
3	1	4	6	2	5	7	8	9

Sudoku # 368

9	5	8	7	3	2	4	1	6
4	1	3	9	8	6	7	5	2
6	7	2	1	5	4	9	8	3
7	9	4	6	2	5	8	3	1
1	2	5	8	7	3	6	9	4
8	3	6	4	9	1	5	2	7
3	6	1	5	4	8	2	7	9
5	4	7	2	1	9	3	6	8
2	8	9	3	6	7	1	4	5

Sudoku # 369

6	8	4	1	3	5	2	7	9
2	7	1	9	4	8	6	3	5
9	3	5	6	7	2	1	4	8
4	2	9	7	5	1	8	6	3
8	1	6	2	9	3	7	5	4
3	5	7	4	8	6	9	2	1
5	6	3	8	2	9	4	1	7
1	4	8	5	6	7	3	9	2
7	9	2	3	1	4	5	8	6

Sudoku # 370

9	1	6	3	5	8	4	7	2
5	7	2	9	1	4	8	3	6
8	3	4	6	7	2	9	1	5
6	2	1	5	9	7	3	4	8
4	9	8	1	2	3	5	6	7
3	5	7	4	8	6	2	9	1
2	4	5	7	3	1	6	8	9
7	6	9	8	4	5	1	2	3
1	8	3	2	6	9	7	5	4

Sudoku # 371

7	6	1	4	5	8	2	9	3
9	2	5	7	6	3	8	4	1
4	8	3	1	9	2	6	5	7
1	7	6	5	4	9	3	2	8
5	3	2	6	8	7	4	1	9
8	9	4	2	3	1	7	6	5
3	5	8	9	2	4	1	7	6
2	1	9	3	7	6	5	8	4
6	4	7	8	1	5	9	3	2

Sudoku # 372

7	4	5	6	3	9	1	2	8
1	3	2	4	5	8	9	6	7
8	9	6	7	1	2	3	5	4
9	2	8	1	7	5	6	4	3
3	5	7	2	4	6	8	9	1
6	1	4	8	9	3	2	7	5
2	7	1	3	6	4	5	8	9
5	6	3	9	8	7	4	1	2
4	8	9	5	2	1	7	3	6

Sudoku # 373

6	8	7	2	1	9	3	4	5
2	3	9	5	6	4	8	7	1
4	1	5	7	8	3	2	6	9
5	6	1	4	2	7	9	8	3
3	7	8	9	5	1	4	2	6
9	2	4	8	3	6	1	5	7
1	9	2	6	7	8	5	3	4
7	5	3	1	4	2	6	9	8
8	4	6	3	9	5	7	1	2

Sudoku # 374

9	6	1	7	8	5	2	4	3
3	8	4	1	9	2	5	6	7
7	2	5	4	3	6	1	9	8
1	3	8	2	7	9	4	5	6
5	4	6	3	1	8	7	2	9
2	9	7	5	6	4	3	8	1
4	7	9	6	2	3	8	1	5
8	1	2	9	5	7	6	3	4
6	5	3	8	4	1	9	7	2

Sudoku # 375

6	3	4	7	1	5	2	9	8
8	1	9	6	4	2	5	3	7
2	7	5	8	9	3	4	1	6
7	8	3	4	5	1	9	6	2
9	4	1	3	2	6	8	7	5
5	2	6	9	8	7	3	4	1
4	9	2	1	7	8	6	5	3
1	6	8	5	3	4	7	2	9
3	5	7	2	6	9	1	8	4

Sudoku # 376

7	8	6	5	3	4	9	2	1
2	3	5	6	1	9	8	4	7
4	1	9	8	7	2	3	5	6
6	2	3	9	4	8	1	7	5
9	5	4	7	6	1	2	3	8
1	7	8	2	5	3	4	6	9
5	4	2	1	8	6	7	9	3
3	6	1	4	9	7	5	8	2
8	9	7	3	2	5	6	1	4

Sudoku # 377

1	3	7	4	2	9	5	8	6
6	8	9	5	3	1	2	4	7
4	5	2	6	8	7	1	9	3
9	4	1	8	5	3	7	6	2
3	6	8	2	7	4	9	1	5
2	7	5	9	1	6	4	3	8
8	1	6	7	4	5	3	2	9
5	2	4	3	9	8	6	7	1
7	9	3	1	6	2	8	5	4

Sudoku # 378

8	9	4	3	7	2	5	6	1
5	6	3	9	8	1	2	4	7
1	2	7	6	4	5	8	3	9
7	5	9	1	6	4	3	8	2
6	8	2	5	9	3	1	7	4
3	4	1	7	2	8	9	5	6
4	7	5	2	3	9	6	1	8
9	1	6	8	5	7	4	2	3
2	3	8	4	1	6	7	9	5

Sudoku # 379

2	1	6	8	3	5	9	7	4
4	3	7	9	1	2	5	6	8
9	8	5	6	7	4	3	1	2
7	6	3	4	2	1	8	5	9
1	2	8	3	5	9	7	4	6
5	4	9	7	6	8	1	2	3
8	5	1	2	4	3	6	9	7
6	9	4	1	8	7	2	3	5
3	7	2	5	9	6	4	8	1

Sudoku # 380

8	7	5	1	4	6	3	2	9
3	9	1	2	7	5	4	8	6
2	4	6	9	3	8	5	1	7
9	2	3	6	5	7	8	4	1
1	5	7	4	8	9	2	6	3
6	8	4	3	2	1	7	9	5
5	3	9	8	1	4	6	7	2
7	1	8	5	6	2	9	3	4
4	6	2	7	9	3	1	5	8

Sudoku # 381

2	6	9	7	8	1	3	4	5
3	1	7	5	4	9	8	6	2
5	4	8	3	6	2	1	9	7
6	9	3	1	7	4	2	5	8
1	5	4	8	2	3	9	7	6
8	7	2	6	9	5	4	3	1
9	3	6	2	1	7	5	8	4
4	8	1	9	5	6	7	2	3
7	2	5	4	3	8	6	1	9

Sudoku # 382

5	1	6	7	2	4	3	8	9
9	7	3	8	1	5	6	2	4
4	2	8	9	3	6	7	5	1
1	8	5	2	9	7	4	6	3
3	4	7	5	6	8	1	9	2
6	9	2	1	4	3	8	7	5
8	6	9	4	5	1	2	3	7
2	3	1	6	7	9	5	4	8
7	5	4	3	8	2	9	1	6

Sudoku # 383

6	8	5	4	2	1	3	7	9
7	9	4	3	5	8	6	1	2
2	3	1	7	9	6	4	8	5
9	5	8	1	6	4	2	3	7
1	7	3	5	8	2	9	6	4
4	2	6	9	3	7	1	5	8
3	1	9	2	7	5	8	4	6
8	4	7	6	1	9	5	2	3
5	6	2	8	4	3	7	9	1

Sudoku # 384

5	9	4	1	7	2	6	3	8
8	3	2	6	4	5	9	1	7
1	7	6	3	9	8	4	5	2
7	1	9	5	8	4	3	2	6
2	8	5	7	6	3	1	9	4
6	4	3	9	2	1	7	8	5
4	6	1	2	5	9	8	7	3
3	5	8	4	1	7	2	6	9
9	2	7	8	3	6	5	4	1

Sudoku # 385

3	4	6	2	5	7	1	9	8
8	2	5	4	1	9	6	7	3
1	9	7	8	3	6	4	2	5
6	8	2	3	4	5	9	1	7
9	1	3	7	6	2	5	8	4
5	7	4	9	8	1	2	3	6
4	6	8	1	2	3	7	5	9
7	5	1	6	9	8	3	4	2
2	3	9	5	7	4	8	6	1

Sudoku # 386

2	9	7	3	5	8	6	1	4
6	8	4	2	7	1	5	9	3
3	5	1	4	6	9	7	2	8
5	4	8	6	1	7	9	3	2
7	2	9	8	4	3	1	6	5
1	3	6	9	2	5	4	8	7
9	7	2	1	8	4	3	5	6
8	1	5	7	3	6	2	4	9
4	6	3	5	9	2	8	7	1

Sudoku # 387

5	3	9	7	2	4	8	6	1
2	6	4	1	8	5	9	7	3
7	8	1	9	6	3	5	4	2
3	5	7	8	4	9	2	1	6
8	9	6	2	1	7	4	3	5
1	4	2	3	5	6	7	8	9
6	1	8	4	9	2	3	5	7
4	2	3	5	7	1	6	9	8
9	7	5	6	3	8	1	2	4

Sudoku # 388

5	8	6	7	4	9	3	2	1
3	1	4	8	2	5	9	6	7
9	2	7	1	6	3	8	5	4
4	6	9	2	8	1	7	3	5
7	5	1	4	3	6	2	8	9
8	3	2	5	9	7	4	1	6
1	4	8	9	5	2	6	7	3
6	9	5	3	7	8	1	4	2
2	7	3	6	1	4	5	9	8

Sudoku # 389

2	7	6	3	8	5	9	1	4
8	3	9	6	4	1	7	5	2
1	5	4	2	9	7	6	3	8
6	1	2	7	3	8	5	4	9
3	4	8	5	1	9	2	6	7
7	9	5	4	6	2	3	8	1
5	8	7	1	2	6	4	9	3
9	6	3	8	7	4	1	2	5
4	2	1	9	5	3	8	7	6

Sudoku # 390

4	2	1	6	7	9	8	3	5
5	7	9	2	8	3	6	4	1
3	6	8	4	1	5	9	2	7
1	8	2	9	6	4	7	5	3
7	5	3	1	2	8	4	6	9
9	4	6	3	5	7	2	1	8
2	1	7	8	3	6	5	9	4
8	3	4	5	9	2	1	7	6
6	9	5	7	4	1	3	8	2

Sudoku # 391

4	5	3	1	2	8	6	9	7
7	2	8	6	9	3	5	1	4
6	1	9	5	7	4	3	8	2
8	3	5	4	1	2	7	6	9
9	4	1	7	5	6	2	3	8
2	7	6	8	3	9	4	5	1
5	6	4	2	8	1	9	7	3
1	9	2	3	6	7	8	4	5
3	8	7	9	4	5	1	2	6

Sudoku # 392

6	7	4	9	8	1	2	5	3
8	5	1	3	2	7	4	6	9
2	9	3	6	5	4	1	8	7
3	1	7	4	6	2	5	9	8
4	6	5	1	9	8	7	3	2
9	2	8	7	3	5	6	4	1
7	4	6	8	1	9	3	2	5
5	3	9	2	7	6	8	1	4
1	8	2	5	4	3	9	7	6

Sudoku # 393

2	7	8	6	5	4	1	9	3
3	6	9	2	7	1	5	8	4
5	4	1	3	8	9	6	2	7
7	3	5	4	2	6	9	1	8
4	9	6	7	1	8	3	5	2
8	1	2	5	9	3	4	7	6
6	2	7	1	3	5	8	4	9
1	8	3	9	4	7	2	6	5
9	5	4	8	6	2	7	3	1

Sudoku # 394

1	4	6	9	3	5	2	8	7
2	8	9	1	6	7	4	3	5
5	7	3	2	8	4	9	1	6
6	1	2	5	9	8	7	4	3
7	9	5	3	4	1	6	2	8
8	3	4	6	7	2	5	9	1
4	2	8	7	5	3	1	6	9
9	5	1	8	2	6	3	7	4
3	6	7	4	1	9	8	5	2

Sudoku # 395

8	5	7	2	3	6	9	1	4
9	6	3	5	4	1	8	2	7
4	2	1	8	7	9	6	3	5
7	3	6	4	1	8	2	5	9
2	9	4	3	5	7	1	6	8
5	1	8	9	6	2	4	7	3
1	4	2	7	9	3	5	8	6
3	8	5	6	2	4	7	9	1
6	7	9	1	8	5	3	4	2

Sudoku # 396

2	3	6	1	9	7	4	5	8
5	4	9	2	3	8	6	1	7
1	7	8	4	6	5	9	2	3
6	1	4	7	8	9	5	3	2
7	5	3	6	4	2	1	8	9
8	9	2	5	1	3	7	6	4
4	8	7	3	5	1	2	9	6
3	2	1	9	7	6	8	4	5
9	6	5	8	2	4	3	7	1

Sudoku # 397

3	8	4	5	7	2	1	9	6
1	6	5	9	8	4	2	7	3
9	2	7	6	3	1	5	4	8
7	9	8	3	4	5	6	2	1
5	1	3	2	6	9	7	8	4
6	4	2	7	1	8	3	5	9
4	7	9	1	2	3	8	6	5
8	3	6	4	5	7	9	1	2
2	5	1	8	9	6	4	3	7

Sudoku # 398

9	7	3	6	1	8	4	5	2
1	6	4	5	7	2	8	9	3
2	5	8	9	3	4	7	6	1
3	4	9	2	8	5	1	7	6
8	1	7	4	6	3	5	2	9
6	2	5	1	9	7	3	4	8
5	3	1	7	2	6	9	8	4
7	9	2	8	4	1	6	3	5
4	8	6	3	5	9	2	1	7

Sudoku # 399

3	2	5	9	8	4	6	7	1
1	8	4	7	5	6	2	3	9
9	7	6	3	2	1	5	8	4
5	6	8	2	3	9	4	1	7
7	9	1	4	6	8	3	2	5
2	4	3	5	1	7	8	9	6
8	3	9	1	4	5	7	6	2
6	5	7	8	9	2	1	4	3
4	1	2	6	7	3	9	5	8

Sudoku # 400

8	6	5	1	2	3	9	4	7
4	1	2	8	7	9	5	3	6
7	9	3	5	6	4	2	1	8
9	2	8	4	3	7	1	6	5
3	4	1	2	5	6	7	8	9
6	5	7	9	8	1	3	2	4
5	3	4	7	1	8	6	9	2
2	8	6	3	9	5	4	7	1
1	7	9	6	4	2	8	5	3

Sudoku # 401

5	3	4	6	2	1	9	8	7
6	1	2	8	9	7	4	3	5
8	7	9	4	5	3	2	1	6
3	4	7	9	1	2	6	5	8
2	6	8	7	4	5	3	9	1
1	9	5	3	6	8	7	2	4
4	8	1	2	7	9	5	6	3
9	5	6	1	3	4	8	7	2
7	2	3	5	8	6	1	4	9

Sudoku # 402

1	9	8	2	4	3	6	7	5
3	5	2	6	1	7	4	8	9
6	7	4	8	5	9	3	1	2
7	2	5	9	6	4	1	3	8
8	4	6	3	2	1	9	5	7
9	1	3	7	8	5	2	4	6
4	3	7	5	9	6	8	2	1
5	8	9	1	3	2	7	6	4
2	6	1	4	7	8	5	9	3

Sudoku # 403

6	1	8	2	5	7	9	3	4
7	2	9	3	6	4	5	1	8
4	3	5	1	9	8	2	7	6
2	7	3	8	1	6	4	9	5
1	8	4	9	3	5	6	2	7
5	9	6	4	7	2	3	8	1
3	6	7	5	8	9	1	4	2
8	4	1	6	2	3	7	5	9
9	5	2	7	4	1	8	6	3

Sudoku # 404

4	5	8	3	1	2	7	6	9
1	7	6	4	5	9	2	8	3
2	9	3	8	6	7	4	5	1
8	2	5	7	3	6	1	9	4
9	6	7	1	2	4	8	3	5
3	4	1	9	8	5	6	7	2
7	8	4	5	9	1	3	2	6
5	3	2	6	4	8	9	1	7
6	1	9	2	7	3	5	4	8

Sudoku # 405

7	8	4	1	9	3	6	5	2
9	1	5	4	2	6	7	3	8
3	6	2	5	8	7	9	4	1
1	2	6	7	4	9	5	8	3
8	3	9	6	5	2	1	7	4
5	4	7	8	3	1	2	9	6
6	5	3	2	7	4	8	1	9
2	9	8	3	1	5	4	6	7
4	7	1	9	6	8	3	2	5

Sudoku # 406

7	8	9	2	5	4	6	1	3
6	2	3	7	8	1	4	9	5
1	5	4	9	6	3	7	8	2
3	1	6	8	4	2	9	5	7
5	4	7	3	9	6	8	2	1
8	9	2	1	7	5	3	6	4
9	6	5	4	2	7	1	3	8
4	3	8	5	1	9	2	7	6
2	7	1	6	3	8	5	4	9

Sudoku # 407

4	2	8	7	9	3	5	1	6
7	3	5	1	6	8	2	9	4
9	1	6	5	2	4	3	8	7
8	5	7	4	1	2	6	3	9
2	9	3	8	7	6	1	4	5
1	6	4	3	5	9	8	7	2
3	7	2	9	8	5	4	6	1
6	4	9	2	3	1	7	5	8
5	8	1	6	4	7	9	2	3

Sudoku # 408

7	5	6	8	4	3	2	1	9
3	2	9	1	6	7	4	5	8
8	4	1	9	2	5	7	6	3
6	1	3	4	7	9	5	8	2
5	7	2	6	1	8	3	9	4
9	8	4	3	5	2	1	7	6
2	9	8	7	3	1	6	4	5
4	3	7	5	9	6	8	2	1
1	6	5	2	8	4	9	3	7

Sudoku # 409

8	2	3	7	4	5	6	9	1
6	9	1	3	2	8	4	5	7
7	5	4	1	6	9	8	2	3
9	3	7	5	8	6	2	1	4
4	6	5	2	3	1	9	7	8
2	1	8	9	7	4	3	6	5
5	4	6	8	1	2	7	3	9
3	8	9	6	5	7	1	4	2
1	7	2	4	9	3	5	8	6

Sudoku # 410

5	2	4	1	9	7	8	6	3
6	1	7	4	8	3	9	5	2
9	3	8	5	6	2	7	4	1
8	4	9	7	2	6	1	3	5
7	5	1	9	3	4	6	2	8
3	6	2	8	5	1	4	9	7
4	8	3	6	7	5	2	1	9
1	9	5	2	4	8	3	7	6
2	7	6	3	1	9	5	8	4

Sudoku # 411

9	4	2	6	3	1	8	5	7
1	8	7	5	4	9	3	6	2
6	3	5	7	8	2	9	4	1
3	2	4	9	6	8	7	1	5
7	6	8	1	2	5	4	3	9
5	9	1	4	7	3	6	2	8
4	1	6	8	5	7	2	9	3
2	7	9	3	1	6	5	8	4
8	5	3	2	9	4	1	7	6

Sudoku # 412

2	7	9	6	5	3	1	8	4
6	5	8	9	1	4	3	2	7
3	1	4	2	8	7	5	9	6
4	6	3	5	2	8	7	1	9
7	8	5	3	9	1	4	6	2
1	9	2	4	7	6	8	3	5
9	4	1	7	3	2	6	5	8
8	2	7	1	6	5	9	4	3
5	3	6	8	4	9	2	7	1

Sudoku # 413

6	9	3	8	1	4	7	5	2
4	1	7	9	2	5	6	3	8
2	5	8	3	6	7	9	1	4
5	8	2	7	3	9	4	6	1
3	4	6	1	5	2	8	9	7
1	7	9	6	4	8	5	2	3
8	6	4	2	9	1	3	7	5
9	2	5	4	7	3	1	8	6
7	3	1	5	8	6	2	4	9

Sudoku # 414

8	6	2	3	4	7	9	1	5
4	9	5	1	6	2	7	8	3
7	3	1	5	8	9	6	2	4
2	5	6	4	1	8	3	7	9
3	1	8	9	7	5	4	6	2
9	7	4	2	3	6	8	5	1
5	2	7	6	9	4	1	3	8
1	8	9	7	2	3	5	4	6
6	4	3	8	5	1	2	9	7

Sudoku # 415

4	2	8	7	9	5	6	3	1
6	1	9	2	3	4	7	8	5
5	3	7	8	1	6	2	4	9
9	4	1	6	2	7	3	5	8
8	7	6	5	4	3	9	1	2
2	5	3	9	8	1	4	7	6
1	8	4	3	6	2	5	9	7
3	6	5	1	7	9	8	2	4
7	9	2	4	5	8	1	6	3

Sudoku # 416

7	1	9	3	4	8	2	5	6
3	2	8	6	7	5	4	1	9
5	6	4	2	9	1	7	8	3
1	4	5	9	8	3	6	2	7
6	9	2	4	1	7	8	3	5
8	3	7	5	2	6	9	4	1
4	8	1	7	5	9	3	6	2
9	5	3	8	6	2	1	7	4
2	7	6	1	3	4	5	9	8

Sudoku # 417

9	1	4	5	2	3	6	8	7
7	2	5	8	1	6	3	9	4
3	8	6	9	7	4	2	1	5
8	6	1	3	4	2	7	5	9
4	3	2	7	9	5	8	6	1
5	7	9	1	6	8	4	3	2
2	4	8	6	5	9	1	7	3
1	5	3	2	8	7	9	4	6
6	9	7	4	3	1	5	2	8

Sudoku # 418

1	9	7	5	3	2	4	6	8
4	5	2	6	8	7	1	9	3
3	6	8	1	9	4	7	5	2
8	2	3	4	5	6	9	1	7
5	1	4	9	7	3	8	2	6
6	7	9	8	2	1	3	4	5
9	3	1	2	6	8	5	7	4
7	4	6	3	1	5	2	8	9
2	8	5	7	4	9	6	3	1

Sudoku # 419

8	9	4	7	3	6	1	2	5
7	1	3	8	5	2	6	9	4
6	2	5	1	4	9	7	3	8
3	4	6	9	8	5	2	1	7
2	7	8	3	6	1	4	5	9
1	5	9	2	7	4	3	8	6
9	3	7	4	2	8	5	6	1
4	6	1	5	9	3	8	7	2
5	8	2	6	1	7	9	4	3

Sudoku # 420

7	9	8	4	3	1	6	5	2
6	1	4	8	5	2	3	7	9
2	3	5	7	6	9	8	4	1
8	7	9	5	2	6	1	3	4
5	4	1	3	9	8	2	6	7
3	6	2	1	4	7	5	9	8
9	8	6	2	7	3	4	1	5
4	2	3	9	1	5	7	8	6
1	5	7	6	8	4	9	2	3

Sudoku # 421

1	9	8	3	4	5	7	2	6
5	6	2	9	1	7	4	3	8
3	4	7	2	8	6	9	5	1
7	1	5	4	3	8	6	9	2
4	3	9	7	6	2	1	8	5
8	2	6	5	9	1	3	7	4
9	7	1	8	5	4	2	6	3
6	8	3	1	2	9	5	4	7
2	5	4	6	7	3	8	1	9

Sudoku # 422

8	9	2	1	3	6	5	4	7
1	5	6	7	4	2	3	8	9
3	4	7	9	8	5	2	6	1
2	6	1	5	7	3	8	9	4
5	3	4	6	9	8	1	7	2
9	7	8	4	2	1	6	3	5
6	8	9	2	5	7	4	1	3
7	2	3	8	1	4	9	5	6
4	1	5	3	6	9	7	2	8

Sudoku # 423

2	1	5	7	8	3	6	9	4
6	9	4	1	2	5	3	8	7
8	7	3	6	4	9	1	2	5
4	6	1	9	5	8	7	3	2
5	2	8	4	3	7	9	1	6
7	3	9	2	1	6	5	4	8
3	8	2	5	6	1	4	7	9
9	4	6	3	7	2	8	5	1
1	5	7	8	9	4	2	6	3

Sudoku # 424

5	3	6	1	9	8	7	2	4
8	9	1	2	4	7	6	5	3
2	4	7	6	3	5	9	8	1
7	2	8	3	5	1	4	9	6
3	5	4	8	6	9	1	7	2
1	6	9	7	2	4	8	3	5
6	7	3	4	8	2	5	1	9
4	8	5	9	1	3	2	6	7
9	1	2	5	7	6	3	4	8

Sudoku # 425

9	1	3	5	7	6	8	2	4
6	5	8	1	4	2	7	3	9
7	4	2	9	3	8	5	6	1
1	8	7	2	9	4	6	5	3
2	6	5	3	8	1	4	9	7
4	3	9	7	6	5	2	1	8
3	2	4	6	1	7	9	8	5
5	7	1	8	2	9	3	4	6
8	9	6	4	5	3	1	7	2

Sudoku # 426

4	6	7	2	5	8	1	3	9
8	5	1	9	4	3	6	2	7
2	3	9	6	7	1	5	8	4
5	9	4	7	1	2	8	6	3
6	1	2	3	8	4	9	7	5
7	8	3	5	6	9	4	1	2
3	2	5	8	9	6	7	4	1
9	4	8	1	3	7	2	5	6
1	7	6	4	2	5	3	9	8

Sudoku # 427

6	1	3	2	7	4	5	9	8
8	9	5	1	3	6	4	2	7
2	4	7	5	8	9	3	6	1
3	2	1	4	5	7	6	8	9
5	6	4	8	9	1	7	3	2
7	8	9	3	6	2	1	4	5
1	5	6	9	2	3	8	7	4
9	7	8	6	4	5	2	1	3
4	3	2	7	1	8	9	5	6

Sudoku # 428

8	6	7	1	9	4	2	5	3
5	9	1	2	7	3	8	6	4
3	2	4	8	6	5	9	1	7
6	3	9	7	4	8	5	2	1
7	1	8	9	5	2	3	4	6
2	4	5	3	1	6	7	9	8
9	5	6	4	8	7	1	3	2
1	7	2	6	3	9	4	8	5
4	8	3	5	2	1	6	7	9

Sudoku # 429

3	5	8	4	6	7	9	1	2
6	4	1	2	3	9	7	5	8
2	7	9	8	1	5	4	6	3
4	1	7	6	9	2	3	8	5
8	3	2	5	7	4	1	9	6
5	9	6	1	8	3	2	7	4
9	2	5	7	4	8	6	3	1
7	6	4	3	5	1	8	2	9
1	8	3	9	2	6	5	4	7

Sudoku # 430

5	3	8	4	9	1	6	7	2
6	7	1	8	5	2	3	4	9
9	2	4	7	6	3	5	8	1
7	5	6	2	3	9	8	1	4
4	1	3	6	8	7	9	2	5
2	8	9	5	1	4	7	6	3
3	6	5	1	2	8	4	9	7
8	4	2	9	7	5	1	3	6
1	9	7	3	4	6	2	5	8

Sudoku # 431

3	1	2	9	4	7	8	5	6
9	8	7	1	6	5	3	4	2
6	4	5	2	3	8	7	9	1
8	2	1	4	9	3	6	7	5
4	5	6	7	8	2	1	3	9
7	9	3	6	5	1	2	8	4
2	7	8	5	1	4	9	6	3
1	6	4	3	7	9	5	2	8
5	3	9	8	2	6	4	1	7

Sudoku # 432

8	7	3	9	5	2	6	4	1
9	6	2	8	4	1	3	7	5
4	5	1	7	3	6	8	2	9
2	9	5	1	8	7	4	6	3
7	1	6	3	2	4	5	9	8
3	4	8	5	6	9	2	1	7
5	2	9	4	1	3	7	8	6
6	8	7	2	9	5	1	3	4
1	3	4	6	7	8	9	5	2

Sudoku # 433

9	7	8	6	1	3	4	5	2
6	3	1	4	2	5	9	8	7
2	4	5	9	8	7	3	6	1
3	5	4	8	7	9	2	1	6
1	2	7	5	4	6	8	3	9
8	6	9	2	3	1	5	7	4
7	8	3	1	9	4	6	2	5
4	1	6	3	5	2	7	9	8
5	9	2	7	6	8	1	4	3

Sudoku # 434

3	1	2	5	6	7	4	9	8
4	6	8	2	3	9	1	7	5
5	9	7	4	1	8	6	2	3
2	7	5	1	9	3	8	6	4
8	3	1	6	4	2	7	5	9
9	4	6	7	8	5	2	3	1
1	8	3	9	2	6	5	4	7
6	5	4	3	7	1	9	8	2
7	2	9	8	5	4	3	1	6

Sudoku # 435

5	7	4	6	8	1	3	2	9
1	9	6	2	4	3	8	7	5
8	2	3	5	7	9	4	6	1
4	6	8	3	9	2	5	1	7
3	1	7	8	5	6	9	4	2
9	5	2	4	1	7	6	3	8
7	8	9	1	6	4	2	5	3
6	3	1	9	2	5	7	8	4
2	4	5	7	3	8	1	9	6

Sudoku # 436

8	3	1	2	9	7	5	4	6
6	9	2	4	8	5	1	7	3
4	5	7	3	6	1	2	8	9
9	8	4	7	3	2	6	1	5
5	7	6	1	4	9	3	2	8
1	2	3	8	5	6	4	9	7
7	4	8	5	2	3	9	6	1
2	6	5	9	1	8	7	3	4
3	1	9	6	7	4	8	5	2

Sudoku # 437

9	6	1	2	5	7	3	4	8
8	7	4	3	1	9	2	5	6
2	5	3	6	4	8	1	7	9
5	1	7	4	2	6	9	8	3
6	3	8	5	9	1	7	2	4
4	9	2	7	8	3	6	1	5
1	8	6	9	7	5	4	3	2
7	2	9	8	3	4	5	6	1
3	4	5	1	6	2	8	9	7

Sudoku # 438

9	5	3	6	7	4	2	1	8
7	6	1	3	2	8	4	5	9
8	2	4	1	9	5	7	3	6
5	3	2	9	1	7	8	6	4
4	9	8	5	6	2	3	7	1
1	7	6	8	4	3	9	2	5
6	8	7	2	5	9	1	4	3
3	4	5	7	8	1	6	9	2
2	1	9	4	3	6	5	8	7

Sudoku # 439

8	4	6	7	9	5	3	2	1
2	1	7	3	8	6	4	9	5
5	9	3	1	2	4	8	6	7
7	3	4	5	1	2	9	8	6
1	8	9	4	6	7	2	5	3
6	2	5	8	3	9	1	7	4
9	6	1	2	7	3	5	4	8
4	7	8	9	5	1	6	3	2
3	5	2	6	4	8	7	1	9

Sudoku # 440

8	3	5	2	4	1	7	6	9
4	7	2	6	9	5	8	1	3
6	9	1	7	8	3	4	5	2
1	2	6	3	7	4	9	8	5
9	5	8	1	6	2	3	4	7
7	4	3	9	5	8	1	2	6
5	8	7	4	3	6	2	9	1
3	1	4	5	2	9	6	7	8
2	6	9	8	1	7	5	3	4

Sudoku # 441

3	2	5	6	4	1	7	8	9
8	7	4	2	3	9	5	6	1
6	9	1	7	5	8	3	2	4
9	3	6	8	2	4	1	7	5
4	8	7	5	1	6	9	3	2
1	5	2	9	7	3	6	4	8
7	6	8	4	9	5	2	1	3
5	4	3	1	6	2	8	9	7
2	1	9	3	8	7	4	5	6

Sudoku # 442

3	4	2	5	6	9	7	1	8
9	7	1	3	2	8	6	5	4
6	5	8	7	1	4	9	2	3
4	6	3	1	9	7	5	8	2
2	1	7	6	8	5	4	3	9
5	8	9	4	3	2	1	6	7
1	9	5	8	7	3	2	4	6
7	3	6	2	4	1	8	9	5
8	2	4	9	5	6	3	7	1

Sudoku # 443

6	2	8	5	9	4	3	1	7
4	1	3	8	2	7	6	5	9
5	9	7	6	1	3	4	2	8
3	7	9	1	4	8	2	6	5
1	8	5	3	6	2	9	7	4
2	6	4	7	5	9	1	8	3
8	3	1	4	7	6	5	9	2
7	5	2	9	3	1	8	4	6
9	4	6	2	8	5	7	3	1

Sudoku # 444

1	3	7	5	4	9	6	8	2
8	9	4	3	2	6	5	7	1
6	5	2	8	7	1	3	9	4
5	4	9	7	6	2	1	3	8
7	2	6	1	8	3	9	4	5
3	8	1	9	5	4	7	2	6
2	7	8	6	9	5	4	1	3
9	6	3	4	1	8	2	5	7
4	1	5	2	3	7	8	6	9

Sudoku # 445

9	8	6	1	4	3	7	2	5
7	3	5	2	8	6	1	9	4
1	2	4	5	7	9	6	3	8
3	5	8	4	2	1	9	7	6
2	9	1	7	6	8	5	4	3
4	6	7	9	3	5	8	1	2
8	7	2	6	9	4	3	5	1
5	4	3	8	1	7	2	6	9
6	1	9	3	5	2	4	8	7

Sudoku # 446

7	9	8	4	3	1	2	5	6
1	4	3	5	6	2	7	8	9
2	6	5	9	7	8	3	4	1
5	7	6	2	8	3	9	1	4
9	3	2	6	1	4	8	7	5
4	8	1	7	9	5	6	2	3
6	5	9	1	2	7	4	3	8
8	2	4	3	5	9	1	6	7
3	1	7	8	4	6	5	9	2

Sudoku # 447

6	2	3	5	7	1	8	9	4
8	7	4	6	2	9	1	5	3
5	9	1	8	3	4	6	2	7
1	3	9	7	5	6	4	8	2
4	8	5	3	9	2	7	6	1
2	6	7	1	4	8	5	3	9
3	5	2	4	8	7	9	1	6
9	4	6	2	1	5	3	7	8
7	1	8	9	6	3	2	4	5

Sudoku # 448

8	5	3	4	9	2	7	1	6
2	9	4	7	6	1	3	8	5
1	6	7	8	5	3	2	4	9
7	3	9	2	4	6	1	5	8
4	8	5	9	1	7	6	3	2
6	1	2	3	8	5	9	7	4
3	4	8	1	2	9	5	6	7
5	2	1	6	7	8	4	9	3
9	7	6	5	3	4	8	2	1

Sudoku # 449

1	9	7	6	4	3	5	2	8
3	2	6	7	8	5	1	4	9
5	4	8	9	2	1	3	7	6
2	3	1	4	7	9	8	6	5
7	5	9	8	1	6	4	3	2
6	8	4	3	5	2	7	9	1
9	6	5	1	3	7	2	8	4
8	7	2	5	6	4	9	1	3
4	1	3	2	9	8	6	5	7

Sudoku # 450

7	9	8	3	5	4	2	1	6
1	2	3	9	6	7	8	5	4
5	4	6	1	2	8	7	9	3
9	1	7	2	3	5	4	6	8
3	5	2	4	8	6	1	7	9
8	6	4	7	9	1	3	2	5
4	8	5	6	1	2	9	3	7
6	3	1	8	7	9	5	4	2
2	7	9	5	4	3	6	8	1

Sudoku # 451

9	8	4	2	7	6	1	5	3
3	1	2	4	8	5	6	9	7
7	5	6	3	9	1	2	8	4
4	6	3	7	1	8	9	2	5
5	9	8	6	4	2	3	7	1
1	2	7	5	3	9	8	4	6
2	4	9	1	6	7	5	3	8
6	7	5	8	2	3	4	1	9
8	3	1	9	5	4	7	6	2

Sudoku # 452

3	7	4	9	1	6	5	2	8
1	9	6	8	5	2	3	7	4
2	5	8	7	4	3	6	9	1
6	1	5	2	9	8	7	4	3
4	8	3	5	6	7	9	1	2
9	2	7	1	3	4	8	6	5
5	4	2	6	8	9	1	3	7
7	6	1	3	2	5	4	8	9
8	3	9	4	7	1	2	5	6

Sudoku # 453

2	6	3	7	4	1	5	9	8
8	1	9	5	6	3	4	7	2
7	4	5	8	2	9	1	6	3
5	3	4	6	9	7	2	8	1
1	9	7	2	8	4	6	3	5
6	2	8	3	1	5	7	4	9
3	7	2	4	5	8	9	1	6
9	8	6	1	7	2	3	5	4
4	5	1	9	3	6	8	2	7

Sudoku # 454

1	7	9	3	8	6	2	4	5
3	2	8	5	9	4	7	1	6
4	5	6	7	2	1	9	3	8
5	9	4	2	6	8	3	7	1
2	8	3	1	4	7	6	5	9
6	1	7	9	5	3	4	8	2
7	6	5	8	3	2	1	9	4
9	4	1	6	7	5	8	2	3
8	3	2	4	1	9	5	6	7

Sudoku # 455

7	9	8	3	4	5	6	1	2
2	6	3	7	8	1	4	5	9
5	1	4	9	2	6	8	3	7
8	2	7	5	3	4	9	6	1
6	5	9	8	1	2	3	7	4
3	4	1	6	9	7	5	2	8
9	7	5	1	6	8	2	4	3
1	8	2	4	5	3	7	9	6
4	3	6	2	7	9	1	8	5

Sudoku # 456

9	6	2	7	5	3	4	1	8
7	4	8	9	2	1	5	6	3
3	1	5	6	8	4	7	2	9
8	7	1	3	4	6	2	9	5
5	2	4	1	9	8	3	7	6
6	3	9	5	7	2	8	4	1
1	8	6	2	3	7	9	5	4
4	9	7	8	1	5	6	3	2
2	5	3	4	6	9	1	8	7

Sudoku # 457

5	9	7	1	8	6	3	4	2
8	3	6	5	2	4	1	7	9
1	2	4	7	9	3	8	5	6
7	8	9	6	4	1	2	3	5
2	4	5	3	7	9	6	1	8
3	6	1	8	5	2	4	9	7
9	5	2	4	1	8	7	6	3
6	1	8	9	3	7	5	2	4
4	7	3	2	6	5	9	8	1

Sudoku # 458

2	1	6	5	7	8	3	9	4
4	5	9	3	2	6	7	8	1
8	7	3	9	4	1	6	2	5
5	2	1	4	9	7	8	3	6
7	3	8	1	6	5	2	4	9
6	9	4	2	8	3	1	5	7
1	8	2	7	5	9	4	6	3
9	4	7	6	3	2	5	1	8
3	6	5	8	1	4	9	7	2

Sudoku # 459

3	2	5	1	9	6	8	4	7
9	6	4	5	7	8	2	1	3
8	7	1	3	2	4	5	9	6
6	9	2	8	3	7	1	5	4
1	4	7	2	5	9	3	6	8
5	3	8	6	4	1	9	7	2
4	1	9	7	8	3	6	2	5
2	8	6	4	1	5	7	3	9
7	5	3	9	6	2	4	8	1

Sudoku # 460

6	3	1	7	8	4	5	2	9
7	2	4	5	3	9	6	8	1
8	9	5	2	1	6	4	3	7
3	4	7	9	6	8	1	5	2
9	8	2	3	5	1	7	4	6
1	5	6	4	2	7	3	9	8
5	1	3	8	7	2	9	6	4
2	6	9	1	4	5	8	7	3
4	7	8	6	9	3	2	1	5

Sudoku # 461

7	5	4	9	2	3	1	6	8
2	1	6	5	8	7	3	4	9
9	8	3	4	1	6	7	2	5
4	9	8	3	7	1	6	5	2
6	7	5	2	9	8	4	3	1
3	2	1	6	4	5	8	9	7
1	6	9	8	5	4	2	7	3
8	4	2	7	3	9	5	1	6
5	3	7	1	6	2	9	8	4

Sudoku # 462

4	5	6	1	7	2	9	3	8
7	2	9	8	4	3	6	5	1
3	8	1	6	5	9	4	2	7
5	4	7	3	6	8	1	9	2
6	3	2	9	1	7	8	4	5
1	9	8	4	2	5	3	7	6
8	6	5	7	3	4	2	1	9
2	1	3	5	9	6	7	8	4
9	7	4	2	8	1	5	6	3

Sudoku # 463

7	3	5	8	2	1	9	4	6
8	6	1	5	4	9	7	3	2
2	4	9	6	7	3	5	8	1
6	1	2	9	5	8	3	7	4
5	8	3	4	1	7	2	6	9
9	7	4	3	6	2	1	5	8
3	9	6	1	8	5	4	2	7
1	2	8	7	3	4	6	9	5
4	5	7	2	9	6	8	1	3

Sudoku # 464

9	6	7	3	5	2	1	4	8
5	3	4	8	6	1	9	2	7
1	2	8	4	9	7	5	6	3
3	9	6	7	2	4	8	5	1
7	4	5	9	1	8	2	3	6
8	1	2	5	3	6	7	9	4
2	7	3	6	8	5	4	1	9
4	5	9	1	7	3	6	8	2
6	8	1	2	4	9	3	7	5

Sudoku # 465

8	2	9	1	7	6	5	4	3
4	6	7	8	5	3	1	2	9
5	3	1	4	9	2	8	6	7
7	1	6	9	4	5	2	3	8
2	4	5	3	6	8	9	7	1
3	9	8	7	2	1	6	5	4
1	5	4	6	8	7	3	9	2
6	7	3	2	1	9	4	8	5
9	8	2	5	3	4	7	1	6

Sudoku # 466

5	6	3	9	8	7	1	4	2
4	1	9	2	3	6	5	7	8
7	8	2	5	1	4	9	6	3
6	4	8	7	2	1	3	5	9
2	9	7	8	5	3	4	1	6
1	3	5	4	6	9	2	8	7
9	2	6	1	7	5	8	3	4
8	7	1	3	4	2	6	9	5
3	5	4	6	9	8	7	2	1

Sudoku # 467

4	3	8	9	1	7	6	5	2
2	6	7	8	4	5	9	3	1
9	1	5	2	6	3	7	4	8
5	9	2	6	7	4	8	1	3
1	8	6	5	3	9	2	7	4
7	4	3	1	2	8	5	9	6
6	2	4	7	5	1	3	8	9
8	5	1	3	9	6	4	2	7
3	7	9	4	8	2	1	6	5

Sudoku # 468

7	4	5	9	8	3	6	2	1
9	8	1	5	2	6	7	3	4
6	2	3	1	7	4	5	8	9
8	7	2	4	1	9	3	5	6
1	9	6	8	3	5	2	4	7
5	3	4	7	6	2	1	9	8
4	5	7	2	9	1	8	6	3
2	6	8	3	4	7	9	1	5
3	1	9	6	5	8	4	7	2

Sudoku # 469

8	7	9	3	1	2	4	5	6
5	3	1	4	8	6	9	7	2
2	4	6	5	7	9	1	3	8
4	8	3	7	2	5	6	9	1
1	9	5	8	6	4	3	2	7
6	2	7	9	3	1	5	8	4
3	1	2	6	9	7	8	4	5
7	5	8	1	4	3	2	6	9
9	6	4	2	5	8	7	1	3

Sudoku # 470

6	3	2	5	9	7	8	4	1
8	7	4	6	2	1	5	3	9
1	9	5	4	8	3	2	6	7
7	2	3	9	4	8	1	5	6
5	4	6	1	3	2	7	9	8
9	1	8	7	5	6	3	2	4
4	6	1	3	7	5	9	8	2
3	8	7	2	6	9	4	1	5
2	5	9	8	1	4	6	7	3

Sudoku # 471

9	1	6	3	8	7	5	2	4
5	3	2	6	9	4	7	1	8
7	8	4	1	2	5	9	3	6
4	7	9	8	3	2	1	6	5
6	2	1	5	4	9	3	8	7
8	5	3	7	1	6	4	9	2
3	9	7	4	6	8	2	5	1
2	6	5	9	7	1	8	4	3
1	4	8	2	5	3	6	7	9

Sudoku # 472

2	3	9	5	1	8	7	4	6
7	5	8	3	6	4	2	9	1
1	6	4	9	2	7	8	5	3
8	7	2	1	5	6	4	3	9
3	4	5	2	8	9	6	1	7
9	1	6	7	4	3	5	8	2
4	2	3	6	9	5	1	7	8
5	9	1	8	7	2	3	6	4
6	8	7	4	3	1	9	2	5

Sudoku # 473

7	4	9	1	8	5	3	2	6
2	8	6	7	9	3	4	5	1
1	5	3	2	4	6	8	7	9
6	7	8	5	1	9	2	3	4
9	3	5	6	2	4	1	8	7
4	1	2	3	7	8	9	6	5
3	6	4	9	5	2	7	1	8
8	2	7	4	6	1	5	9	3
5	9	1	8	3	7	6	4	2

Sudoku # 474

2	4	3	7	8	9	6	5	1
5	7	8	4	1	6	3	9	2
1	9	6	3	2	5	8	7	4
7	5	1	8	9	4	2	3	6
9	3	2	5	6	1	4	8	7
8	6	4	2	7	3	5	1	9
3	1	5	6	4	7	9	2	8
6	8	7	9	3	2	1	4	5
4	2	9	1	5	8	7	6	3

Sudoku # 475

3	9	1	5	8	6	4	2	7
7	5	6	4	1	2	9	8	3
4	2	8	3	7	9	5	1	6
6	4	5	9	3	8	1	7	2
8	3	7	2	4	1	6	5	9
9	1	2	6	5	7	8	3	4
2	6	3	8	9	5	7	4	1
5	7	4	1	6	3	2	9	8
1	8	9	7	2	4	3	6	5

Sudoku # 476

6	9	5	1	4	8	3	2	7
1	7	3	2	5	6	9	4	8
4	2	8	3	7	9	6	5	1
7	6	2	9	1	3	5	8	4
8	4	1	5	6	2	7	3	9
3	5	9	4	8	7	2	1	6
9	3	6	8	2	4	1	7	5
2	1	4	7	9	5	8	6	3
5	8	7	6	3	1	4	9	2

Sudoku # 477

6	4	1	8	5	7	3	9	2
8	2	5	6	9	3	4	1	7
9	7	3	2	4	1	5	8	6
1	6	9	4	8	2	7	5	3
2	8	4	7	3	5	9	6	1
3	5	7	1	6	9	8	2	4
5	9	6	3	2	4	1	7	8
7	3	2	9	1	8	6	4	5
4	1	8	5	7	6	2	3	9

Sudoku # 478

8	6	7	3	9	2	1	5	4
2	1	3	8	4	5	9	6	7
4	9	5	1	6	7	8	2	3
6	7	9	4	1	8	2	3	5
3	8	1	5	2	6	7	4	9
5	2	4	7	3	9	6	1	8
1	5	8	2	7	3	4	9	6
7	4	6	9	5	1	3	8	2
9	3	2	6	8	4	5	7	1

Sudoku # 479

6	2	4	3	1	5	7	8	9
7	1	3	6	9	8	5	2	4
8	5	9	4	7	2	6	3	1
4	3	5	1	2	7	8	9	6
9	6	1	5	8	3	2	4	7
2	7	8	9	6	4	1	5	3
3	9	6	2	5	1	4	7	8
5	4	7	8	3	6	9	1	2
1	8	2	7	4	9	3	6	5

Sudoku # 480

5	6	3	7	4	2	8	1	9
1	4	8	3	6	9	2	7	5
7	2	9	8	5	1	6	3	4
2	1	4	5	3	7	9	6	8
9	5	7	6	1	8	3	4	2
3	8	6	2	9	4	1	5	7
6	7	2	1	8	5	4	9	3
4	3	5	9	2	6	7	8	1
8	9	1	4	7	3	5	2	6

Sudoku # 481

3	5	7	8	6	2	9	4	1
1	6	8	9	7	4	5	2	3
9	4	2	3	1	5	6	8	7
4	7	6	5	2	1	3	9	8
8	3	1	4	9	7	2	5	6
2	9	5	6	8	3	7	1	4
6	8	4	7	5	9	1	3	2
5	2	3	1	4	6	8	7	9
7	1	9	2	3	8	4	6	5

Sudoku # 482

6	9	8	1	3	2	5	4	7
5	2	3	8	7	4	1	9	6
4	7	1	5	9	6	3	8	2
7	5	2	3	1	9	8	6	4
1	8	9	4	6	7	2	3	5
3	4	6	2	8	5	7	1	9
2	1	4	9	5	3	6	7	8
9	3	7	6	2	8	4	5	1
8	6	5	7	4	1	9	2	3

Sudoku # 483

8	2	3	7	6	4	1	9	5
4	6	1	2	5	9	7	3	8
7	5	9	1	8	3	6	2	4
3	9	5	8	1	7	4	6	2
2	7	4	3	9	6	5	8	1
1	8	6	5	4	2	3	7	9
5	1	7	6	2	8	9	4	3
6	4	2	9	3	5	8	1	7
9	3	8	4	7	1	2	5	6

Sudoku # 484

9	3	6	2	1	4	7	5	8
4	7	5	8	6	3	1	2	9
1	8	2	9	5	7	4	6	3
5	2	7	6	8	9	3	4	1
3	9	8	1	4	2	6	7	5
6	1	4	7	3	5	9	8	2
8	4	1	3	2	6	5	9	7
2	6	9	5	7	1	8	3	4
7	5	3	4	9	8	2	1	6

Sudoku # 485

9	3	7	4	2	1	5	8	6
8	1	5	7	3	6	2	4	9
2	4	6	9	5	8	3	7	1
4	5	9	2	8	7	1	6	3
3	6	1	5	9	4	8	2	7
7	8	2	1	6	3	4	9	5
1	7	3	8	4	9	6	5	2
6	2	4	3	7	5	9	1	8
5	9	8	6	1	2	7	3	4

Sudoku # 486

4	9	7	2	3	5	8	6	1
2	8	5	9	1	6	7	3	4
6	1	3	8	4	7	2	5	9
8	6	4	5	7	9	3	1	2
5	2	1	3	6	4	9	8	7
3	7	9	1	2	8	6	4	5
1	4	8	7	9	3	5	2	6
9	3	6	4	5	2	1	7	8
7	5	2	6	8	1	4	9	3

Sudoku # 487

4	8	3	7	6	1	5	9	2
9	2	5	3	4	8	6	7	1
6	7	1	9	5	2	3	8	4
8	9	4	5	1	7	2	3	6
3	1	2	6	9	4	8	5	7
5	6	7	8	2	3	1	4	9
1	5	9	4	8	6	7	2	3
7	4	6	2	3	5	9	1	8
2	3	8	1	7	9	4	6	5

Sudoku # 488

4	3	5	2	7	6	1	9	8
9	2	1	3	4	8	7	6	5
7	6	8	1	9	5	4	3	2
5	8	2	6	3	1	9	4	7
1	7	9	8	5	4	6	2	3
6	4	3	9	2	7	8	5	1
3	9	6	7	1	2	5	8	4
2	1	4	5	8	9	3	7	6
8	5	7	4	6	3	2	1	9

Sudoku # 489

9	7	8	1	5	3	2	4	6
5	4	6	2	8	9	7	3	1
1	3	2	6	7	4	5	9	8
6	1	3	7	9	8	4	2	5
8	2	7	5	4	1	9	6	3
4	5	9	3	2	6	1	8	7
3	8	5	9	1	2	6	7	4
7	9	4	8	6	5	3	1	2
2	6	1	4	3	7	8	5	9

Sudoku # 490

6	4	3	2	7	9	8	1	5
2	8	5	6	1	4	3	7	9
7	1	9	8	5	3	4	2	6
1	2	8	4	6	7	5	9	3
4	9	7	3	8	5	2	6	1
3	5	6	9	2	1	7	8	4
5	3	2	1	9	8	6	4	7
8	7	1	5	4	6	9	3	2
9	6	4	7	3	2	1	5	8

Sudoku # 491

1	5	6	8	9	2	7	3	4
2	9	8	7	4	3	1	5	6
3	4	7	1	6	5	8	9	2
5	2	9	3	1	8	4	6	7
4	8	1	6	5	7	9	2	3
6	7	3	9	2	4	5	1	8
9	6	4	2	8	1	3	7	5
8	3	2	5	7	9	6	4	1
7	1	5	4	3	6	2	8	9

Sudoku # 492

2	8	9	1	3	7	4	5	6
4	3	7	6	8	5	9	1	2
5	1	6	2	4	9	7	3	8
7	9	5	3	6	1	2	8	4
3	2	8	9	7	4	5	6	1
6	4	1	8	5	2	3	9	7
1	5	4	7	9	6	8	2	3
8	7	2	5	1	3	6	4	9
9	6	3	4	2	8	1	7	5

Sudoku # 493

7	2	3	1	6	9	8	5	4
9	8	1	2	5	4	6	3	7
4	5	6	8	3	7	1	9	2
5	3	7	6	4	8	9	2	1
2	1	9	5	7	3	4	8	6
6	4	8	9	2	1	3	7	5
8	6	5	4	9	2	7	1	3
1	7	2	3	8	6	5	4	9
3	9	4	7	1	5	2	6	8

Sudoku # 494

2	4	7	3	6	5	8	1	9
8	1	9	2	4	7	6	3	5
6	5	3	1	8	9	4	7	2
4	2	5	7	9	3	1	6	8
1	7	6	5	2	8	3	9	4
3	9	8	6	1	4	5	2	7
7	3	1	4	5	2	9	8	6
5	8	2	9	3	6	7	4	1
9	6	4	8	7	1	2	5	3

Sudoku # 495

2	4	6	9	8	3	1	5	7
1	9	8	6	7	5	2	4	3
7	3	5	4	2	1	9	6	8
5	6	4	2	9	7	8	3	1
8	2	9	3	1	6	4	7	5
3	7	1	5	4	8	6	2	9
9	5	3	8	6	2	7	1	4
6	8	7	1	5	4	3	9	2
4	1	2	7	3	9	5	8	6

Sudoku # 496

7	1	5	6	9	3	2	8	4
2	3	8	7	4	1	6	5	9
6	4	9	8	2	5	3	1	7
9	2	7	5	1	6	4	3	8
1	5	6	3	8	4	7	9	2
3	8	4	2	7	9	5	6	1
4	7	3	9	6	8	1	2	5
5	9	1	4	3	2	8	7	6
8	6	2	1	5	7	9	4	3

Sudoku # 497

8	7	6	5	3	1	9	4	2
9	4	3	6	2	8	7	5	1
1	2	5	4	7	9	8	6	3
3	9	4	7	6	2	5	1	8
2	5	8	3	1	4	6	7	9
6	1	7	9	8	5	2	3	4
7	8	2	1	4	6	3	9	5
4	3	9	8	5	7	1	2	6
5	6	1	2	9	3	4	8	7

Sudoku # 498

1	5	8	2	9	7	3	6	4
9	2	3	1	4	6	8	5	7
4	7	6	5	8	3	2	1	9
8	9	2	3	6	5	7	4	1
3	1	4	7	2	9	5	8	6
5	6	7	8	1	4	9	3	2
2	3	1	4	7	8	6	9	5
7	8	9	6	5	1	4	2	3
6	4	5	9	3	2	1	7	8

Sudoku # 499

3	2	1	6	8	5	9	4	7
7	5	8	2	9	4	3	1	6
6	4	9	3	1	7	8	5	2
9	3	4	8	5	2	6	7	1
1	6	5	4	7	3	2	8	9
2	8	7	1	6	9	4	3	5
4	9	2	7	3	1	5	6	8
5	7	6	9	4	8	1	2	3
8	1	3	5	2	6	7	9	4

Sudoku # 500

9	3	2	7	6	8	4	5	1
8	1	4	3	2	5	6	9	7
5	6	7	9	4	1	2	8	3
6	8	1	4	5	3	9	7	2
3	7	9	6	1	2	5	4	8
4	2	5	8	9	7	1	3	6
2	4	8	1	3	9	7	6	5
1	9	3	5	7	6	8	2	4
7	5	6	2	8	4	3	1	9

THIS BOOK IS A TOTAL BLAST, AND IF YOU WANT TO SHOW SOME LOVE, SCAN THE QR CODE AND MAKE A CONTRIBUTION TO OUR TIP JAR!

SCAN QR CODE TO VIEW MORE BOOKS.

ELMSLEIGH DESIGNS
PASSION FOR DESIGN

www.ingramcontent.com/pod-product-compliance
Lightning Source LLC
Chambersburg PA
CBHW070909290526
45795CB00001B/264

* 9 7 9 8 8 7 9 1 1 8 7 8 0 *